AF812455

Synthese Library

Studies in Epistemology, Logic, Methodology, and Philosophy of Science

Volume 517

Editor-in-Chief

Otávio Bueno, Department of Philosophy, University of Miami, Coral Gables, USA

Editorial Board Members

Berit Brogaard, University of Miami, Coral Gables, USA

Steven French, University of Leeds, Leeds, UK

Catarina Dutilh Novaes, VU Amsterdam, Amsterdam, The Netherlands

Darrell P. Rowbottom, Department of Philosophy, Lingnan University, Tuen Mun, Hong Kong

Emma Ruttkamp, Department of Philosophy, University of South Africa, Pretoria, South Africa

Kristie Miller, Department of Philosophy, Centre for Time, University of Sydney, Sydney, Australia

The aim of *Synthese Library* is to provide a forum for the best current work in the methodology and philosophy of science and in epistemology, all broadly understood. A wide variety of different approaches have traditionally been represented in the Library, and every effort is made to maintain this variety, not for its own sake, but because we believe that there are many fruitful and illuminating approaches to the philosophy of science and related disciplines.

Special attention is paid to methodological studies which illustrate the interplay of empirical and philosophical viewpoints and to contributions to the formal (logical, set-theoretical, mathematical, information-theoretical, decision-theoretical, etc.) methodology of empirical sciences. Likewise, the applications of logical methods to epistemology as well as philosophically and methodologically relevant studies in logic are strongly encouraged. The emphasis on logic will be tempered by interest in the psychological, historical, and sociological aspects of science. In addition to monographs *Synthese Library* publishes thematically unified anthologies and edited volumes with a well-defined topical focus inside the aim and scope of the book series. The contributions in the volumes are expected to be focused and structurally organized in accordance with the central theme(s), and should be tied together by an extensive editorial introduction or set of introductions if the volume is divided into parts. An extensive bibliography and index are mandatory.

Lars Arthur Tump

On Mathematical Explanations of Empirical Phenomena

Indispensability, Number Theory, and Mathematical Counterfactual Dependence

Springer

Lars Arthur Tump
Institute of Philosophy, Centre for Logic and Philosophy of Science
KU Leuven
Leuven, Belgium

ISSN 0166-6991 ISSN 2542-8292 (electronic)
Synthese Library
ISBN 978-3-031-98859-2 ISBN 978-3-031-98860-8 (eBook)
https://doi.org/10.1007/978-3-031-98860-8

© The Editor(s) (if applicable) and The Author(s), under exclusive license to Springer Nature Switzerland AG 2025

This work is subject to copyright. All rights are solely and exclusively licensed by the Publisher, whether the whole or part of the material is concerned, specifically the rights of translation, reprinting, reuse of illustrations, recitation, broadcasting, reproduction on microfilms or in any other physical way, and transmission or information storage and retrieval, electronic adaptation, computer software, or by similar or dissimilar methodology now known or hereafter developed.
The use of general descriptive names, registered names, trademarks, service marks, etc. in this publication does not imply, even in the absence of a specific statement, that such names are exempt from the relevant protective laws and regulations and therefore free for general use.
The publisher, the authors and the editors are safe to assume that the advice and information in this book are believed to be true and accurate at the date of publication. Neither the publisher nor the authors or the editors give a warranty, expressed or implied, with respect to the material contained herein or for any errors or omissions that may have been made. The publisher remains neutral with regard to jurisdictional claims in published maps and institutional affiliations.

This Springer imprint is published by the registered company Springer Nature Switzerland AG
The registered company address is: Gewerbestrasse 11, 6330 Cham, Switzerland

If disposing of this product, please recycle the paper.

Acknowledgments

Many people have been of support during the process leading up to writing this book. I would like to take the opportunity to express my gratitude. First and foremost, I would like to thank my former PhD supervisor Jan Heylen at the Institute of Philosophy, KU Leuven. I have thoroughly enjoyed our working together, both on a personal and professional level. Our discussions, no matter the topic, were always of interest. Besides, my work has greatly benefited from his support. Our joint work, as referenced here below, is testimony of this. I would also like to thank the people of the Centre for Logic and Philosophy of Science at KU Leuven. Working environments are created by people, and I've experienced a stimulating and friendly one. I would like to mention Sylvia Wenmackers, Grant Ramsey, Jan Opsomer, Felipe Morales Carbonell, Leander Vignero, Bendik Aaby, Alejandro Gordillo García, Kristine Grigoryan, Pieter Thyssen, Lorenz Demey, Olivier Lemeire, Andreas De Block, Stefaan Cuypers. I would also like to thank Otávio Bueno, as this book would probably not have happened had he not been the external member during my PhD defense. Some of the research carried out was supported by a PhD scholarship from the Humanities and Social Sciences Group, KU Leuven. I thank them for that. This book is based on my doctoral thesis, of which parts were already published in various journals. Below a specification:

- Chapters 2 and 4 are based on the paper "The enhanced indispensability argument, the circularity problem, and the interpretability strategy" (Heylen, Jan and Lars Arthur Tump 2021, online 2019), which is published in *Synthese* 198(4), 3033–3045. https://doi.org/10.1007/s11229-019-02263-0. © Springer Nature BV 2019. Used with permission.
- Chapters 7 and 8 are based on the paper "Mathematical counterfactuals with number-theoretic antecedents and extra-mathematical explanation" (Tump, Lars Arthur 2021), which is published in *Logique et Analyse* 254, 191–213. https://doi.org/10.2143/LEA.254.0.3289657. © Peeters Publishers 2021. Used with permission.

Competing Interests The author has no competing interests to declare that are relevant to the content of this manuscript.

Contents

Chapter 1
Introduction

Abstract In this introductory chapter, the subject of mathematical explanation of empirical phenomena is positioned in relation to the topics of scientific explanation and mathematical explanation. This serves as the stepping stone towards the main matter of the book. Part one, The Indispensability of Mathematics, pertains to the role of mathematical entities in mathematical explanations of empirical phenomena. Part two, Mathematical Counterfactual Dependence, pertains to the topic of mathematical explanation within the framework of a counterfactual theory of explanation.

The main topic of this book is mathematical explanation of empirical phenomena. To provide explanations is one of the important roles of scientific theories. A scientific explanation may tell us, for example, why the planets orbit the sun the way they do. Or it may tell us why matter expands and contracts given changes in temperature. As such, scientific explanations contribute to our knowledge and understanding of the world around us. To be sure, providing explanations is not the only important role of scientific theories. Providing us with descriptions of the phenomena around us is another, as is providing us with the ability to make predictions about future events.

A scientific theory does not necessarily need to accommodate all these roles. Take for instance Kepler's three laws of planetary motion. These are ingenious descriptions of the planetary orbits, based on observational data, but they do not explain the trajectorial paths of the planets. In contrast, Newton's mechanics *did* explain the orbit of the planets. It did so by introducing the idea of a gravitational force, by which means it was possible to account for the dynamics of the celestial bodies (Taylor, 2005, p. 91). On Newton's account, the gravitational force is the fundamental concept from which the trajectorial paths of the planets are explained. In contrast to Kepler's laws of planetary motion, Newton's theory thus surpasses the merely descriptive.

The substantial use of mathematics in the sciences, most notably perhaps in physics, but also in chemistry, biology, and other sciences, raises the question what this says about the role and status of the specific mathematics that is employed

© The Author(s), under exclusive license to Springer Nature Switzerland AG 2025

L. A. Tump, *On Mathematical Explanations of Empirical Phenomena*,
Synthese Library 517, https://doi.org/10.1007/978-3-031-98860-8_1

in the various theories. In our case, the focus will be on the use of mathematics in scientific explanations. The mathematics that is employed can range from sometimes relatively simple calculations to oftentimes very complex models. In either case, however, one can ask whether mathematics does (part of) the work in the explanation of the phenomenon in question. And if it does so, how would such mathematical explanations work, given the abstract nature of mathematics and the concrete nature of the phenomenon? It seems that such questions cannot be answered independently of the broader discussions about the relation between the status of mathematics on the one hand and that of the scientific phenomena on the other. In fact, any account of mathematical explanation appears to be in need of some account of the relation between mathematics and reality, that is, of an account that relates the abstract to the concrete. With an eye to this, a subsequent question would be what the role of mathematics in scientific explanations could say about the ontological status of the particular mathematics that is used in those explanations, assuming, amongst other things, that it indeed were to contribute to the explanation.

This introductory chapter presents the topic of explanation. Section 1.1 introduces explanations in science. Section 1.2 then discusses explanation within mathematics, and what it could mean for mathematics to be explanatory. In Sect. 1.3, which is concerned with the topic of mathematical explanations in science, the interrelation between the topics of the previous sections is discussed. Having set the stage, the general outline of the remainder of the book, which is divided up into two parts and concluded by an epilogue, is presented in Sect. 1.4.

Let me emphasize right off the bet that, for reasons of scope, the focus in this book will be on those kind of theories of explanation that fit the deductive nomological theory of explanation in Part I, and the counterfactual theory of explanation in Part II. This choice results quite naturally from the specific examples of scientific explanations that are being discussed in the current literature, and specifically with regard to the example that is used as the entry point for the discussions in this book. It is of course recognized that there are other accounts of explanation, such as statistical accounts or unification accounts. However, the main goal of the book is not so much to argue for a specific account of mathematical explanation, but rather to investigate what the consequences are if certain arguments are to be made in light of questions of explanation and ontology. At this point, since the matter of scope has been brought up, it might also be instructive to mention that the context in which this research takes place is broadly empiricist in nature.

1.1 Explanations in Science

Before we are in a position to delve into the topic of mathematical explanations of empirical phenomena, let alone the details of some of the arguments related to it, we must have some idea of what scientific explanations are typically considered to be. Generally speaking, explanations can be thought of as collections of statements intended to show why something is the case. A scientific explanation, for instance, sets out to answer the question 'why?', in relation to some phenomenon. In their famous article 'Studies in the Logic of Explanation', Hempel and Oppenheim (1948) give as an example the workings of a mercury thermometer. What needs to be explained, the explanandum, can be captured by the following question: why does the mercury temporarily drop and then rapidly rise when the thermometer is immersed in hot water? According to them, the answer to this question, the explanans, is the following:

> [t]he increase in temperature affects at first only the glass tube of the thermometer; it expands and thus provides a larger space for the mercury inside, whose surface therefore drops. As soon as by heat conduction the rise in temperature reaches the mercury, however, the latter expands, and as its coefficient of expansion is considerably larger than that of glass, a rise of the mercury level results (Hempel & Oppenheim, 1948, pp. 135–136).

Historically, much down to the influence of German idealism, 'why-questions' were closely tied to metaphysically laden ideas about understanding. And so much so, that certain physicists, like Ernst Mach or Pierre Duhem, said that science was not in the business of asking such why questions at all (Curd & Cover, 1998, p. 767).[1] Rather, the role of science was only to represent or classify empirical laws (Duhem, 1906a, p. 19).

This changed when the deductive-nomological (DN) theory of explanation was formulated by the very same (Hempel & Oppenheim, 1948). From then onward, when 'why-questions' were asked, what was generally meant was for that which was to be explained to be put in a framework of empirical laws (Carnap, 1966, p. 678). The DN-theory of explanation provides such a framework. Specifically, it considers explanations to be deductively valid arguments in which the explanandum is the conclusion of the argument and must follow logically from the explanans, and in which at least one of the premises is a general law. For this reason, it is also known as the covering law thesis, because it requires any explanation to be an argument involving at least one general law. An important reason for requiring the involvement of a general law in the argument is to exclude so-called 'accidental generalizations'. These are generalizations that are actually true, but not necessarily so. This would be the case, for instance, if the description of a certain regularity were to be restricted in some form or other, but which could have been different had the scope been different. A couple of well-known examples are the following (Van Fraassen, 1989, p. 27):

> All solid spheres of gold (Au) have a diameter of less than one mile.

With regard to this example, Van Fraassen observes that, although the sentence evaluates to true on earth (because there is no solid sphere of gold larger than one mile on earth), there is no scientific reason to exclude the existence of such golden spheres in general. The diameter of one mile does not seem to pose a physical

[1] The use of why-question for the introduction of the idea of explanation here is not intended to mean that a theory of explanation is also one of why-questions, as Van Fraassen (1980, p. 134) argued. See Sandborg (1998) for a criticism of such an account. However, it is clear that the idea of explanation and answers to why-questions are closely related, as we will see later on.

limitation preventing golden spheres to be larger than it. There could, in principle, exist a golden sphere larger in diameter on another planet. More so, even if no planet were to harbour any, this would still not pose a principled scientific reason to exclude the possibility of such a solid sphere to exist.

In contrast, the following proposition does seem to reference a physical limitation, and could as such be up for consideration for the status of law:

> All solid spheres of enriched uranium (U235) have a diameter of less than one mile.

The physical limitation here is related to the critical mass of uranium. Because of its critical mass, it is not possible for there to exist a solid sphere of uranium with a diameter larger than one mile (clearly, only a fraction of it would be possible). Interestingly, as can be seen in the formulation of the evaluation of both sentences, the role of counterfactuals is important here. The (rudimentary) explanations make use of alternative possibilities or impossibilities, which are signalled by the use of 'could' or 'could not'. The role of counterfactual statements will be briefly touched upon later in this section and in more detail in Part II.

According to the DN-theory, an large class of scientific explanations exhibit this covering law pattern. Take for instance the earlier mentioned example of the thermometer. There, a combination of initial conditions and general empirical laws are used in the explanans to show why the explanandum must be the case. The laws that are used in the explanans are those pertaining to the thermic expansion of matter, in casu mercury and glass. Given some initial conditions describing the thermometer and its state just before its immersion in hot water, the explanandum should logically follow from the initial conditions and the empirical laws referenced in the explanans. The conclusion of the explanatory argument should amount to a description of the phenomenon, and the observational statements describing the drop and rise of the mercury in the thermometer should match the deductively begotten explanandum. As will be addressed below, the law, or laws, involved in the deduction must be essential in deriving the conclusion.

Hempel and Oppenheim (1948, p. 137) sum up four conditions for the adequacy for DN-explanations, three of them being logical in nature (R1–R3), and one of them empirical (R4):

R1 The explanandum must be a logical consequence of the explanans.
R2 The explanans must contain general laws, and these must be essential for the derivation of the explanandum.
R3 The explanans must have empirical content, that is, the explanans must be, at least in principle, testable by experience and observation.[2]
R4 The sentences in the explanans must be true.

Parenthetically, it is interesting to see that this way of capturing explanations is quite similar to that of capturing predictions, in the sense that the structure of both

[2] Because it is assumed that the explanans contains at least some empirical content, this requirement follows, strictly speaking, from R1.

is very similar. In the case of an explanation, it is about what *was* to be expected, as an explanation is after the fact. When changing grammatical tense, in the case of a prediction it is about what *is* to be expected, as a prediction is before the fact (Strevens, 2006, p. 4).

Research on the DN-theory of explanation has been, and still is, extensive. And criticisms of it abound. Our goal is not to take a position in this debate. However, the DN-theory of explanation is of interest here due to the fact that the example of a mathematical explanation of an empirical phenomenon discussed in Part I does seem to fit it quite well. But before continuing, let us briefly discuss some issues surrounding the theory.

One of the problems of the DN-theory of explanation is that not all explanations do seem to fit it well. Elliptically formulated explanations, i.e. explanations where certain facts or laws are left out, do strictly speaking not qualify as explanations in this framework. However, as Hempel (1965, p. 415) correctly observes, this kind of incompleteness is quite unproblematic. The missing facts and laws are taken for granted in the elliptical explanation and can easily be filled in when required. The resulting, and complete, argument will then fit the DN-theory again.

Partial explanations, on the other hand, are more problematic. In fact, they are not covered by the DN-theory at all. The reason for this is that a partial explanation is not a deductive argument. The conclusion cannot follow from the premises, even after filling in all the available details that might have been left out elliptically. As the name suggests, inevitably some gaps remain in a partial explanation. Therefore, according to the DN-theory, partial explanations do not qualify as explanations. It is easily observed that partial explanations violate requirement R1. However, what often *does* seem to follow from a partial explanation is a more general conclusion, whereby the initially intended conclusion is one of the possible instances of the application of a more general phenomenon which is explained. Thus, even though the specifics of the conclusion cannot be derived from the explanans, for example when trying to answer the question why an apple changes colour from green to red when ripening, a partial explanation can be seen as an instance of the explanation of the more general phenomenon of the changing of color of apples when ripening. It is this more general phenomenon that is actually entailed, and not the specific instance of changing colour from green to red (Curd & Cover, 1998, p. 773). Reiterating the main takeaway, a partial explanation is not a deductively complete argument and therefore not a DN-explanation (Hempel, 1965, pp. 416–417).

Another important class of explanations, this time based on inductive reasoning, is also excluded by the DN-theory of explanation. The most prominent examples of accounts of inductive explanation are perhaps the probabilistic theories of explanation. Hempel himself tried to fill the lacuna of the DN-theory of explanation by developing the Inductive-Statistical (IS) theory. Its structure is similar to the DN-theory in that it still exhibits the traits of the covering law thesis. The difference is that the universal physical law which was required in the DN-theory is replaced by a probabilistic law. This means that the explanandum is not entailed by the premises of the explanans anymore, but rather assigns a high probability to it (Hempel, 1965, pp. 378–379). The explanans in an IS-explanation is thus that which is statistically relevant to something to be explained.[3]

Although the DN-theory excludes certain types of explanation, in some respects it is too permissive. There are well-known and extensively discussed examples of this. Probably the most notable examples are related to the problems of irrelevancy and symmetry. Irrelevancy occurs when an explanation contains a law,[4] which turns out to be nonessential for the phenomenon to be explained. Nevertheless, the argument itself is deductively sound. For example:

> This sample of table salt dissolves in water, for it has had a dissolving spell cast upon it, and all samples of table salt that have had dissolving spells cast upon them dissolve in water (Salmon, 1970, pp. 33–34).

Even though the argument follows the pattern of the DN-theory of explanation, it is clear that hexing is irrelevant for deriving the conclusion. After all, salt dissolves in water anyway, being hexed or not.

The problem of symmetry occurs, again, when an argument follows the pattern of the DN-theory, but the inverse of the argument would do so too. For example:

> A flagpole of height H, casts a shadow of length S. Given the law that light travels in straight lines, and the elevation of the sun, θ, we can deduce the length of the shadow from the height of the flagpole ... using the equation $H = S\,tan\,\theta$. But given that same equation, we could have ... deduced the height of the flagpole from the length of its shadow (Curd & Cover, 1998, pp. 778–779).

It is generally accepted that both directions of the inference constitute predictions, but it is generally not accepted that the inverse direction is an explanation. Given the intuition that explanations are not symmetrical, the symmetry exhibited in the tangential equation is problematic for the DN-theory of explanation.

To remedy the above mentioned problems, explanations require a relevance relation between explanans and explanandum. An obvious candidate is to make the relevance relation a causal relation (Kitcher, 1989, p. 419). The absence of causal notions in the DN-theory of explanation is, after all, one of the reasons why these problems occur. One of the advantages of causal relations[5] is that they are not considered to be symmetrical. Causes are to precede effects. This means that if a causal relation holds between the law in the explanans and the event or phenomenon in the explanandum, then the argument itself is not symmetrical as a result.

The problem of irrelevancy is more tenacious, because a causal relevance relation per se might not offer a solution. Take the example of the hexed salt. A causal

[3] There are other probabilistic accounts of explanation, e.g. Salmon's (1970) account of statistical relevance, or Railton's (1978) deductive-nomological model of probabilistic explanation, amongst others.

[4] It is not immediately clear whether generalizations such as used in these counterexamples qualify as laws. To judge so, however, requires a theory of what laws are. For the purpose of illustration, it suffices to treat statements that are generalizations without exceptions as laws or at least law-like.

[5] These causal relations are sometimes cashed out in terms of probabilistic relevance, see for instance Salmon (1970).

account could rule out the hexing of the salt as being relevant to the dissolution of it in water, but whether it would do so depends on the notion of causality that is employed. If causation is thought of in terms of regularities (of succession of objects) without exception, akin to the view Hume and many in the empiricist tradition held, then it would of course not help out. The regularity that each time salt is hexed it dissolves in water is exceptionless, and therefore not excluded if this definition is employed. Alternatively, if the notion of causality is thought of as resembling something close to what is used in the more fundamental sciences, similar to Salmon's (1984) suggestion that the total of all the causal influences determines an explanation, then the problem of irrelevance comes up in a different guise. How to determine which of the causal influences is relevant to some event? If, say, a baseball hits a window, and all causal influences count, then the shouting of the players during the impact of the baseball should also be explanatory relevant to the breaking of the window (Strevens, 2006, p. 17). This, however, appears to be contestable at the very least. Generally, the intuition is that only the impact of the baseball, and not the shouting, is relevant for the explanation of the breaking of the window. A similar observation was already made by Lewis (1986), when he stated that some parts of a causal history are not important when explaining the occurrence of a certain event, or for explaining what caused the event. But they belong to the totality of its causal history nonetheless.

A different take on causality came with the development of counterfactual theories of causation, most notably the one developed by Lewis (1973a). Taking inspiration from Hume, Lewis focused on the second definition of causation from said philosopher. Apart from defining causation as regularities without exception, Hume also defined causation in what we nowadays would call a counterfactual way: "we may define a cause ... where, if the first object had not been, the second never had existed (Hume, 1748, sect. VII). Lewis (1979, p. 459) also thought causes are to precede effects, and that in most cases[6] the causal asymmetry is due to the temporal asymmetry of the events. The moment at which the cause occurs (t_C) is earlier in time than the moment at which the effect occurs (t_E).

This intuitive idea about the role of causation as a difference maker is then put in the framework of possible-world semantics. A counterfactual will then be a statement with a false antecedent, stating a possible alternative to the actual situation, and an accordingly modified consequent (Lewis, 1973a, p. 558). Given certain counterfactual dependencies, whatever they may be, the goal is to ultimately determine the relevance of some causal relation for an explanation. This way, the counterfactual account intends to settle the question of relevance. Of course, by augmenting the DN-theory with a theory of causation, the notions of explanation and causality become tied up substantially, and perhaps against the spirit of the DN-theory.

[6] Lewis (1979, pp. 457–458) considers counterfactuals to harbour some inherent vagueness, and leaves open the possibility that there might be situations in which asymmetry does not hold, for instance in scenario's with time machines, or when counterfactually stating something about a time before the Big Bang (although it might not make sense from the viewpoint of physics).

The idea of thinking in terms of counterfactual dependencies is at the core of Woodward's (2003) interventionist account of explanation, in which what he calls 'what-if-things-had-been-different information' plays a key role (Woodward, 2003, p. 14). His proposal is to view explanation as showing systematic patterns of counterfactual dependence between the explanans and the explanandum:

> one ought to be able to associate with any successful explanation a hypothetical or counterfactual experiment that shows us that and how manipulation of the factors mentioned in the explanation ... would be a way of manipulating or altering the phenomenon explained (Woodward, 2003, p. 11).

More technically, both the explanans and the explanandum contain variables, X and Y respectively, which can take different values x and y. The explanans contains a generalization G, such as the tangential equation in the case of the flagpole, which relates the variables X and Y. If after an intervention, for instance a change in the height of the flagpole, X takes a value x' different from some initial condition x, then the generalization should correctly describe the changes to the value of Y, resulting in y' and with $y' \neq y$. That is, the value of the variable expressing the length of the shadow should change according to the changes to the variable expressing the height of the flagpole (Woodward, 2003, p. 203).

According to Woodward, there is a close connection between the counterfactual analysis of causation on the one hand, and scientific explanation on the other. Both exploit the relevant modalities. That is, the modal explanatory connections Woodward has in mind are (predominantly) causal connections. Additionally, when considering this characterization of explanation, it is clear that the notion of 'intervention' plays an important role, hence the name of the theory. Importantly, an intervention is interdefined with causation: an intervention is on a so-called 'intervention variable', I, which causes the variable X to take a different value, which in turn has an effect on the value of Y. It is important to notice that the intervention variable, once a certain value for I is obtained, ensures that X causally depends on I. In other words, other possible causal influences are 'switched off' (Woodward, 2003, p. 98).

It must be noted however, that the door is left open for counterfactual dependencies to not necessarily be equated with causal dependencies (see Saatsi & Pexton, 2013, p. 614 or Reutlinger, 2016, p. 738). Specifically in those cases where a counterfactual dependence does not give an answer to what would (causally) happen in case of an intervention. For instance when mathematics is concerned (Woodward, 2003, p. 221). The particulars of such accounts will be detailed in Chap. 6.

1.2 Explanations in Mathematics

Mathematical explanations of mathematical facts are explanations aimed at questions internal to mathematics, and in which the explanatory aspect comes from mathematics itself. In other words, the 'why-question' as to why some mathematical fact is the case, analogous to the why-question asked in a scientific explanation, is answered by the mathematics invoked with respect to this mathematical fact. Mostly, such explanations are proofs which, besides establishing the truth of some theorem, also explain why the theorem holds. In that respect an explanatory proof is supposed to give additional information that goes beyond just showing the logical consequences. Let us look at a simple example of what could be considered an explanatory proof according to Colyvan (2012). Let's say we ask ourselves the following question: why is 2 the only even prime number? An answer could be given by the following short reductio argument:

> assume ... that there is another even prime, $p > 2$. Since p is even, it can be divided by 2 so it can be written as $p = 2q$ for some integer $q > 1$. But this means that p is composite, contradicting the original assumption that p is prime (Colyvan, 2012, p. 80).

The explanatory aspect of this proof comes from the additional information about even numbers, specifically their property of being divisible by 2 (without remainder). This additional information is then used to give insight in why there cannot be other even prime numbers than 2.

Of course, not all proofs are considered to be explanatory. Take for example proofs by exhaustion, or case proofs. These proofs recognize the fact that there are different cases, which jointly exhaust all the available possibilities. It is then proven that some result holds for each and every case (Colyvan, 2012, p. 91). Such proofs are generally not considered to be explanatory, because they merely demonstrate by brute force that something is the case by going over all the available possibilities. They show *that* something is the case, but they don't show *why* something is the case.

What *does* makes a proof explanatory, according to Steiner (1978a, p. 143), is when it references a characterizing property. And importantly, the result of the proof depends on the use of this property. In the example just given above, the characterizing property would thus be being divisible by 2 for all even numbers, but not for the prime numbers. Except for the number 2, which is both divisible by 2 and a prime number. This characterizing property allows for the extra information that the proof provides. Additionally, he states that:

> [i]t must be evident ... that if we substitute in the proof a different object of the same domain, the theorem collapses; more, we should be able to see as we vary the object how the theorem changes in response. In effect, then, [mathematical] explanation is not simply a relation between a proof and a theorem; rather, a relation between an array of proofs and an array of theorems ... (Steiner, 1978a, p. 143).

To show what is meant by this, Steiner gives the following example to contrast a non-explanatory proof with an explanatory proof of the following theorem (Steiner, 1978a, p. 136):

$$\sum_{n=1}^{n} n = 1 + 2 + 3 + \ldots + n = \frac{n(n+1)}{2}$$

The first proof of the theorem he presents is a proof by mathematical induction, and is meant to illustrate a non-explanatory proof. It is first shown that the theorem holds for n, which can be checked easily, and subsequently it is shown that the theorem also holds for $n + 1$:

$$\sum_{n=1}^{n}(n+1) = \sum_{n=1}^{n}(n) + (n+1) = \frac{n(n+1)}{2} + \frac{2(n+1)}{2} = \frac{(n+1)(n+2)}{2}$$

The second proof of the theorem makes use of the symmetry of addition, and is meant to illustrate an explanatory proof:

$$\begin{array}{ccccccccc} 1 & + & 2 & + & 3 & + \ldots + & n \\ n & + & (n-1) & + & (n-2) & + \ldots + & 1 \\ \hline (n+1) & + & (n+1) & + & (n+1) & + \ldots + & (n+1) \end{array}$$

Here, the initial sequence is duplicated in reverse order, thereby making use of the commutativity of addition, creating an otherwise identical second sequence. These sequences are then added up positionally, resulting in n times (depending on the length of the sequence) the result of the positional addition $n + 1$. And because the initial sequence was duplicated, the result needs to be divided by 2.

Steiner's observation is that the property of symmetry in the sum of the natural numbers, $1 + 2 + \ldots + n$, gives us insight in the truth of the theorem. It is the characterizing property in the proof. What is more, varying the object in the proof, i.e. the sequence in the theorem used in the example, will 'deform' the proof and result in a different theorem (Steiner, 1978a, p. 143). A simple illustration from Gijsbers (2017, p. 50) shows this nicely. Take the sequence $1+3+5+\ldots+(2n-1)$. Using the same method as used in the explanatory proof above, i.e. making use of the commutativity of addition, we get the following:

$$\begin{array}{ccccccccc} 1 & + & 3 & + & 5 & + \ldots + & (2n-1) \\ (2n-1) & + & (2n-3) & + & (2n-5) & + \ldots + & 1 \\ \hline 2n & + & 2n & + & 2n & + \ldots + & 2n \end{array}$$

The same reasoning applies here: we have n times $2n$, which equals $2n^2$. Dividing the result by 2, given the duplication, gives us n^2. What is important to notice here, is that the explanation effectively surpasses a specific proof. In reference to the characterizing property, namely the commutativity of addition, the object is varied. That is, a different sequence is used. As such, a single proof might not be considered explanatory. Each single proof is, so to speak, embedded in a general 'proof-idea', within which the object can be varied (Steiner, 1978a, p. 147). It is by contrasting the different proofs which fit the proof-idea that the explanatory aspect is brought out. Each individual proof is part of the explanation, because they are legitimate instances of the substitution of the object in the proof-idea, but the proofs do not individually answer the 'why-question' (Weber & Verhoeven, 2002, p. 300).

Although the idea of a characterizing property and the idea of the varying, or 'deforming', of a proof are interrelated in the theory of Steiner (1978a), they can be considered separately as well. This is relevant, because the idea of a characterizing property is not without criticism. Some argue that the notion itself is rather vaguely defined. For instance, it is not entirely clear what counts as relevant for determining the domain of a property unique to a structure (Resnik & Kushner, 1987; Gijsbers, 2017). Some argue that there are counterexamples to the idea of a characterizing property, and that there are explanatory proofs not making use of it. See for instance Hafner and Mancosu (2005).

There are also some concerns with regard to the idea of varying the object in a proof-idea. First, how to determine what *the* object to vary is in a proof, given that there are several syntactically distinct objects, i.e. the three sequences in the example. Of these three, two are of course extensionally identical, as a result of the commutativity of addition. Second, it is often the case that the specific presentation of an object in a proof is of importance, and, specifically, the properties related to these descriptions. For instance, when comparing the two different sequences $1+2+3+\ldots+n$ and $1+3+5+\ldots+(2n-1)$. Is the fact that the sequences have different (not necessarily characterizing) properties, such as adding 1 vs. adding 2, not more important than the fact that the object has been varied? It could be argued that this is what might explain the difference in theorem.

It is yet to be mentioned that whether a proof is considered explanatory or not is contentious as well. For example, there might be some who think the inductive proof is in fact explanatory, for instance a mathematical logician. More specifically, whether or not a proof is considered explanatory might depend on the (sub)domain some individual is part of. For instance, said mathematical logician might find some proofs explanatory that a topologist might not, and vice versa (Resnik & Kushner, 1987, p. 146).

Notwithstanding these difficulties, there seems to be a sense in which some proofs are ruled out as explanatory from the get go, such as proofs by exhaustion, whereas for others this is not the case, such as Steiner intended to show with his discussion of the topic. Thus, recapitulating, the idea of a characterizing property is specifically about a proof failing if some entity or structure is substituted for another entity or structure without that characterizing property. For instance if the symmetry property of addition were to be absent in Steiner's example of an explanatory proof. The idea of varying the object within a proof-idea, and the corresponding observations as to what that amounts to, is about substituting some object in the proof with a different object, which results in a different (but related) theorem. It is noteworthy that the proof by induction which Steiner considered non-explanatory does not exhibit either of those. For now, given the scope of this book, and in particular the topic of counterfactual theories of explanation in Part II, the emphasis will be on the idea of varying the object.

1.3 Mathematical Explanations in Science

The success of causal theories of explanation, as introduced in Sect. 1.1, led to scientific explanations primarily being thought of in terms of causality. By and large, this became the dominant view: for something to be an explanation is to identify or describe a cause (Reutlinger, 2017; Pincock, 2015). Of course, the idea that the class of explanations is larger than the class of causal explanations was already well known. Take theories of statistical explanation, for instance. These are not causal theories. Also, theories that emphasize unification are not necessarily causal in nature. And, as mentioned earlier, even the DN -theory of explanation itself is also not inherently causal in nature.

The past couple of decades or so, interest in non-causal explanations has taken a different turn. It was observed that many of the non-causal explanations appear to rely on mathematics doing some of the explanatory work. Thereby, the mathematics is thought to be (at least partly) constitutive of the explanation of the empirical phenomenon in question. Mathematics is of course ubiquitous in science. A brief survey through the sciences makes this abundantly clear. However, mathematics is considered to be causally inert. That is to say, it is atemporal, aspatial, and inefficacious. Therefore, the reasoning goes, the explanations containing elements of mathematics cannot themselves be (fully) causal explanations. This is, of course, not to say that causal explanations are of less importance, or even less prevalent. Recent work on causal explanation by Strevens (2008) is testimony of that. What it does mean, is that this different turn opens up a new perspective on explanations of empirical phenomena.

The topic of mathematical explanation in science is contentious. Not only is there no agreement on whether mathematics actually does explain, or does some of the explaining of, empirical phenomena, but there is also no agreement on how mathematics explains, if it would do so. This has led to the study of mathematical explanations of empirical phenomena, also called extra-mathematical explanations, itself. The term 'extra-mathematical explanation' is used in contrast to 'intra-mathematical explanation', which are mathematical explanations of mathematical facts, such as mentioned in the previous section.

As alluded to above, one of the questions is what it is that makes an explanation of an empirical phenomenon *distinctively* mathematical? It cannot just mean that the explanans quantifies over mathematical entities, such as numbers, functions, or sets. Because this way, many explanations which generally are considered to be ordinary scientific explanations would qualify as distinctively mathematical (Lange, 2017, p. 11). Take for instance the example of the mercury thermometer by Hempel and Oppenheim again. This (causal) explanation mentions the difference in the coefficient of expansion between glass and mercury. This coefficient of expansion is given by a mathematical formula. The formula typically states the relation between the rate of expansion or contraction of some substance given an increase or drop in temperature. This means that the explanation of the temporary drop and rapid rise of the mercury would quantify over mathematical entities and therefore, given the scope of the quantifiers, would also have to be considered to be distinctively mathematical in that sense. The point is that this does not seem to do justice to the intuition that there are relevant differences between ordinary scientific explanations, although they might employ mathematics, even essentially so, and other scientific explanations that seem to hinge on mathematics for their explanations in quite a different way (Lange, 2012, p. 8).

When we recall the idea of the characterizing property of an explanatory mathematical proof, by way of example, then it is clear that in the case of the thermometer the explanation does not hinge on such a mathematical property, but rather on empirical matters of fact, albeit described by the mathematics. In case of a distinctively mathematical explanation of an empirical phenomenon, apart from the fact that, basically by definition, the mathematics figures essentially in the mathematical explanation, the result of the explanation must also depend (in part) on the characterizing property.

Some, notably Steiner (1978b), therefore think that a mathematical explanation in science requires a mathematical explanation in mathematics. Considered this way, an extra-mathematical explanation is said to supervene on the explanatory mathematical proof. The mathematical part is related to the physical part by the relevant assumptions of the physical theory. These assumptions function as the bridge between the mathematics and the physical theory (Steiner, 1978b, p. 19). The mathematical explanation in mathematics emerges when, as it were, stripping off the empirical parts of the extra-mathematical explanation. This is reminiscent of a discussion on the topic by Einstein (1921) when he speaks about the relation between axiomatic geometry and practical geometry. The former being the purely mathematical variety, whereas the latter is the variety used in physics, such as, for instance, in his theories of relativity. Practical geometry is obtained by a coordinative definition relating real, as in physically existing, objects to "the empty conceptual schemata of axiomatic geometry" (Einstein, 1921, p. 235). Similarly, Lyon and Colyvan (2008, p. 241) also hold the view that (part of) the extra-mathematical explanation comes from the mathematical explanatory proofs or conjectures themselves, given some appropriate bridge principles linking the mathematical explanation in pure mathematics to its scientific application.

But this is not universally accepted. Baker (2015, pp. 236–239), for instance, argues specifically against the idea that mathematical explanations in science require a mathematical explanation in mathematics. Part of his reasons can be found in the difficulty of finding a generally accepted definition of what an explanatory proof is, as mentioned in Sect. 1.2. But also in the fact that a proof might not be explanatory in itself, yet fulfills a crucial role in the explanation of a scientific phenomenon. All that is needed then, is that the result of the mathematical proof features essentially in the scientific explanation, be the mathematical proof itself explanatory or not. It needs to be seen, case by case, whether an explanatory proof is needed for an extra-mathematical explanation or not.

A related issue is concerned with explanatoriness itself. Within the class of explanatory mathematical proofs, some proofs are considered to be more explanatory than others. For instance, when a proof is more general, it is often considered to be more explanatory. Similarly, in cases of mathematical explanations in science, theoretical virtues also play an important role. For example, when comparing (possible) mathematical explanations in science, the notion of generalization is often mentioned as a virtue because it allows for inferences to be drawn that exceed some specific instance. This is also called ‘topic generality’, because the explanation itself has little to do with the topic of the specific instance, other than that the specific instance shares some relevant aspects with other specific instances (Baker, 2017b, pp. 200–201). Knowles and Saatsi (2019, p. 1125) give a toy example of a topic general explanation:

> [a] philosopher tries to divide twenty-three distinct ideas evenly among three sections, but keeps failing. Why? Because:
>
> (A) There is no integer n such that $23/3 = n$.

It is easy to see that the explanation applies to all distinct things, not just distinct ideas. No collection of 23 distinct things can be divided up evenly in 3 groups without ending up with a remainder. The fact that this can occur should come as no surprise, for, after all, the (positive) integers are not closed under division.

Another form of generality is so-called ‘scope generality’, which is a generality that allows the explanation to answer questions beyond the scope of a singular instance, or beyond a range of instances bounded in scope. Scope generality can also be found in the above toy example. Not only can the explanation deal with situations in which one tries to divide 23 things evenly over 3 groups of things, but it can also deal with situations where a whole range of a different number of things and a different number of groups is the case. Number theory, as it were, guarantees this possibility.

Thus, unification is often considered to contribute to explanations (Kitcher, 1989). Take an explanation that is topic general. It is considered to be unifying in the sense that it has the ability to connect seemingly unrelated topics, such as 23 ideas and 23 strawberries. It needs to be mentioned here that some claim that it is also possible for unification to have the opposite effect on explanations, because unificatory elements can be introduced that do allow for unification in a certain respect, but at the same time fail to explain the workings of these introduced elements.

A good example of this is Newtonian gravitational theory. Morrison (2000, p. 4) notices that although the introduction of the notion of *force* explained the movement of the planets, the newly introduced force itself was completely mysterious. In fact, Leibniz and Huygens, who were contemporaries of Newton, rejected the theory because it did not fit the, at the time dominant, mechanistic world view in which all such phenomena are explained in terms of the motions of, in principle observable, corpuscles (Dijksterhuis, 1950, pp. 479–478). This disagreement amongst world views, by the way, highlights an important aspect of possible answers to why-questions. The context within which these are asked can differ significantly, and as a result what counts as a possible scientific explanation too. In that sense, a why-question implicitly fixes what constitutes a possible answer in relation to the question it asks, including the conceptual framework that is employed (Sandborg,

1998, pp. 621–622). From a mechanistic point of view, action at a distance would not be acceptable as part of an explanation. Nor would it later be in a relativistic world view. In any case, we see that attempts at unifying the terrestrial and celestial have not always taken explanatory power into account.

Coming from a different angle, but nevertheless closely related to the issue of the question of the nature of extra-mathematical explanation, is the question how mathematics relates to science. Not solely in terms of, for example, its use with regard to idealizations, but also in terms of its use in models in general. Generally speaking, models of theories can be seen as sets of sentences whereby each sentence of the theory is true (Halvorson, 2019, p. 22).[7] Many of these scientific models make use of mathematical sentences, or mixed mathematical and physical sentences (this will be further discussed in Sect. 2.1). But for it to be possible for there to be extra-mathematical explanations, it appears that the mathematics needs to at least capture some structural features of the phenomenon it explains:

> [s]trictly speaking, a mathematical description, model, structure, theory, or whatever, cannot serve as an explanation of a non-mathematical event without an account of the relationship between mathematics per se and scientific reality per se. Without such an account, it is not clear how "scientific explanations" succeed in explaining anything (Shapiro, 1983, p. 525).

Historically, this question was thought to be settled. No effort was made to separate mathematics and science, as can be seen in the work of Descartes, Galilei, Huygens, or Newton (Maddy, 2008, p. 17). This could be done, because at the time it was believed that mathematics described nature correctly. But this changed in the nineteenth century, when substantial developments occurred in mathematics, for example with regard to n-dimensional geometry, negative numbers, or complex numbers. This meant that mathematics became separated from, and in fact went beyond, its application in the sciences. Mathematics was no longer a description of nature (Kline, 1972, p. 1030). Interestingly though, many a development that initially went beyond the applicability in the sciences have later found their way back into scientific practice in the twentieth century. For example, complex numbers were found to be of use in quantum mechanics.

Given the above, the issue concerning mathematical explanations in science ultimately requires an account of the relation between mathematics, and mathematical modeling, on the one hand, and the natural phenomena on the other, such as we have seen with Einstein's coordinative definition linking practical geometry and axiomatic geometry. But, as Saatsi (2011, p. 145) remarks, the fact that through mathematics we can gain knowledge of the physical world does in itself not entail that the mathematics which is employed plays an explanatory role. The explanatory

[7] The question about what theories are will not be addressed here. Suffice it to say that it is a large topic and beyond the present scope. Take for instance the question whether theories should be seen as sets of sentences or sets of models. Or the question whether theories are to be seen as structured, i.e. besides consisting of the collection of elements (for instance sets or sentences), also consisting of the (inferential) relations between the elements. For a discussion on the topic, see Halvorson (2019, p. 277).

role needs to be established independently. Baron (2016), for instance, argues that the explanatory load that mathematics carries might best be understood in terms of difference-making, specifically counterfactually. Difference-making will be addressed in both Parts I and II of this book, whereas the role of counterfactuals will be discussed in Part II.

1.4 Outline

Having introduced the idea of mathematical explanation of empirical phenomena, we are now in a position to proceed to the main matter. The remainder of the book consists of two parts. Part I pertains to a specific ontological argument in the philosophy of mathematics called the 'enhanced indispensability argument'. This argument is about the role of mathematical entities in mathematical explanations of empirical phenomena. It implies that, if mathematical entities play an indispensable explanatory role, one is justified to infer the existence of these mathematical entities. The main concern of Part I is what the limitations internal to the argument are for providing support for an account of mathematical realism.

Chapter 2 will introduce the argument and provide the relevant context for the remainder of the first part. Chapter 3 will address some, but not all, of the many objections raised against it. The discussed objections are relevant either because they relate directly to arguments presented in the subsequent chapters, or because they address general assumptions of the enhanced indispensability argument.

Chapter 4 introduces a further problem internal to the argument, building on an objection pertaining to circularity. To prevent circularity, a strategy of paraphrasing away the mathematical content in the explanandum is needed. This, however, results in a change in context between the mathematical explanans and paraphrased explanandum. To bridge the difference in context, a principle called 'Hume's Principle' is suggested, which could be added to the explanation. Interestingly, it appears to be so that the question of its circularity depends on the outcome of the debate on the analyticity of Hume's Principle. If Hume's Principle turns out to be analytical, then the argument remains circular. In the case where the principle turns out to be not analytical, the second part of the chapter offers an alternative non-mathematical explanation making use of a theory of concatenation and relative interpretability. It is argued that the problem of circularity comes up again if proponents of the enhanced indispensability argument argue that the mathematical explanation has more strength. Chapter 5 summarizes the results from the first part of the book.

Part II discusses the topic of mathematical explanation within the framework of a counterfactual theory of explanation, and discusses a particular suggestion for capturing number theoretic explanatoriness of an empirical phenomenon counterfactually, namely by allowing for a change in necessary properties. The main question is what the effects of such changes are, and whether we still speak about the same numbers in both the factual and the counterfactual case. This is particularly relevant, because an explanation is post hoc. That is, the empirical content which needs explaining is a given.

Chapter 6 sets the stage by first introducing a way in which mathematical explanation could be embedded in a counterfactual theory of explanation. Second, given the necessary nature of mathematics, relevant aspects of discussions on counterpossibles, i.e. counterfactuals with impossible antecedents, are discussed. Chapter 7 then introduces the suggested number theoretic counterpossible about the natural numbers, and raises some initial worries pertaining to contradictions that might arise. It also presents a couple of possible ways in which a solution can be sought, exemplified by making use of Peano Arithmetic. The upshot is that if one redefines an arithmetical function such as multiplication so as to accommodate certain specific changes, then the resulting arithmetic is not isomorphic to the standard model of arithmetic anymore. Alternatively, if one changes the same arithmetical function via the axioms of Peano Arithmetic, then there are downward changes which result in a change in the successor function. The idea of succession is often considered as fundamental in our understanding of the natural numbers.

Chapter 8 discusses both options in context of discussions on the intended models of arithmetic, computability, and de re propositional attitudes and the importance of the successor function. The conclusion is that the ramification are such that the resulting structure might not be that of the natural numbers anymore. The chapter concludes with some further ramifications where, similar to the difference in context between the explanans and explanandum mentioned in Part I, the meaning of the antecedent of the counterpossible now differs from the empirically begotten content of the consequent. Contrary to the invocation of Hume's Principle in Part I, it is not clear what principle could fulfill a similar role here. Chapter 9 provides a summary of Part II.

Part I
The Indispensability of Mathematics

Chapter 2
The Enhanced Indispensability Argument

Abstract This chapter introduces the enhanced indispensability argument. Given the criticism of confirmation holism in the original Quine-Putnam indispensability argument, in conjunction with the idea that indispensability is not considered to be sufficient for drawing ontological conclusions, the enhanced indispensability argument offers an explanatory version of the indispensability argument and states that the entities involved should be indispensable in the right way, namely indispensable for explanations.

The enhanced indispensability argument (EIA from here on), and the backdrop against which it is formulated, will be introduced in this chapter. To that end, the original Quine-Putnam indispensability argument will be presented and discussed in Sect. 2.1. Emphasis will be put on objections raised to the idea of the confirmation holism in this argument. This is relevant because, contrary to the original argument, the enhanced version does not intend rely on confirmation holism. In the enhanced argument, the entities under consideration for ontological commitment are limited to those that play an indispensable role in scientific explanations. After the general context has been provided, the enhanced argument itself is then introduced in Sect. 2.2. The enhanced argument in turn serves as the lead up to a concrete proposal of an extra-mathematical explanation, and a discussion of some of the potential difficulties any putative example might face. These are the topics of Sect. 2.3. In Sect. 2.4 we are then in the position to introduce the concrete proposal. Finally, Sect. 2.5 summarizes the relevant aspects with a view to the coming chapters, which deal with a selection of objections to the EIA in the literature, and the specific problem of the circularity of the argument.

2.1 The Quine-Putnam Indispensability Argument

One of the great controversies in the philosophy of mathematics is about the existence of mathematical entities. Mathematical platonists affirm the existence of mathematical entities, whereas mathematical nominalists deny it.

© The Author(s), under exclusive license to Springer Nature Switzerland AG 2025
L. A. Tump, *On Mathematical Explanations of Empirical Phenomena*, Synthese Library 517, https://doi.org/10.1007/978-3-031-98860-8_2

A major argument for mathematical platonism is the so-called 'Quine-Putnam indispensability argument', of which the core ideas are said to trace back to Quine (1948) and Putnam (1971), and from which authors the argument takes its name.[1]

The key theses of the argument can be stated as follows (Ketland 1998, p. 70):

Thesis A Science is platonistic (Quine 1948).

Thesis B Platonistic science is indispensable (Putnam 1971).

Over the years, the Quine-Putnam indispensability argument, and its supporting theses A and B, have spurred a lot of debate, e.g. (Field 1980; Burgess & Rosen 1997; Melia 2000; Azzouni 2004; Leng 2010), to name a selection.

Support for Thesis A consists in the fact that theories belonging to natural science are formulated in *mathematicized languages*, i.e. languages that contain mathematical vocabulary, such as constants, functions and relation symbols. Also, these theories contain mixed vocabulary, i.e. function symbols and relation symbols that are combined with some mathematical terms and some physical terms. These in turn are used in mixed formulas. Take for instance the well known second law of Newton,[2]

$$\mathbf{F} = m\mathbf{a}$$

where, when using the International System of Units (SI), **F** is the force on an object at a certain time in newton (N), *m* the mass of an object in kilogram (kg), and **a** the acceleration of an object in meter per second squared (m/s2). This is a mixed formula, because underlying the shorthand formulation of the law are functions relating physical terms to mathematical terms, in this case point masses[3] to real numbers. Ketland (1998, p. 49) restates the law as follows, to bring out the complete picture:

> for any point mass x and any instant t, there are real numbers X_1, X_2, X_3 such that X_1 *is the mass-in-kg* of x at time t, X_2 *is the acceleration-in-*m/s^2 of x at time t, X_3 *is the total force-in-N* on x at time t, and $X_3 = X_1 \times X_2$.

[1] The argument should not be taken as a presentation of either Quine's or Putnam's arguments in relation to platonism. Rather, the what is now called 'Quine-Putnam indispensability argument' is inspired by ideas found in their work. For an interesting discussion on the relation between the two namesakes and the argument, see Liggins (2008). In his later work, Putnam (2006, p. 183) explicitly distances himself from the argument as intended by Quine: "[m]y 'indispensability' argument (as opposed to Quine's) was an argument for the objectivity of mathematics in a realist sense—that is, for the idea that mathematical truth must not be identified with provability. Quine's indispensability argument was an argument for 'reluctant Platonism', which he himself characterized as accepting the existence of 'intangible objects' (numbers and sets)".

[2] See for instance sections 1.4 and 1.5 of Taylor (2005).

[3] Point particles, such as the point masses mentioned here, are what Taylor (2005, p. 13) calls 'convenient fiction'. The role of idealizations will be discussed shortly.

Moreover, not only are the theories belonging to natural science highly mathematicized, the theories also contain *existentially quantified sentences*, i.e. sentences of the form $\exists x\phi$ with bound variables in ϕ that range over both the physical as well as the mathematical entities. Take for instance Newton's third law,

$$\mathbf{F}_{AB} = -\mathbf{F}_{BA}$$

which states that the force exerted by one body on another body is equal to, but in opposite direction of, the force exerted by the latter to the former. Reformulating the law, with F as the predicate for the vector force, brings out the existential quantification:

$$\forall F \forall x \forall t \exists y (F(x, t) = -F(y, t))$$

According to Quine's criterion of ontological commitment, this means that those theories from natural science are ontologically committed to the existence of mathematical entities. The criterion of ontological commitment is probably best known as the slogan "to be is to be a value of a bound variable" (Quine 1953, p. 15). This captures the, for Quine, intimate relation between ontological commitment and quantification.

To see this intimate relation, let us look at a sentence ϕ containing a predicate P and a variable x. Taken together they form the open formula Px, meaning 'x is P'. The open formula is true in case the predicate P is true of an object a. Adding the existential quantifier to the open formula, to form the sentence $\exists x Px$, and thereby making the variable bound, assures that there has to be at least one object making the sentence true. If a theory contains such an existentially quantified sentence, then there must be an object making the sentence true if the theory is to be true.[4]

Quine's criterion itself does however not say anything about *what* entities exist. It only states what entities one is committed to given the acceptance of a certain theory. The dialectics of this is important to keep in mind. As Quine states it:

> I look to variables and quantification for evidence as to what a theory says that there is, not for evidence as to what there is (Quine 1960b, p. 243).

For Quine, one reason to look to variables and quantification as evidence for ontological commitment was the problem of non-referring terms (Quine 1948). Quine took cue from Russell's theory of descriptions (Russell 1905), which provided a method for analyzing certain sentences in which an existence (or non-existence) claim is made by means of so-called 'definite descriptions'. Definite descriptions are descriptive names such as 'the author of Waverly' or 'the present King of France'. At first glance, a sentence in which the first definite description occurs should be assigned the truth value TRUE, because there is an author of the

[4] It is clear that this way of paraphrasing presupposes classical logic. For a insightful discussion on this, and possible consequences in meta-ontological discussions of it, see Smid (2020).

book, and the sentence in which the second occurs the truth value FALSE, because there is presently no king of France. However, assigning the truth value FALSE to the sentence containing the non-referring definite description appears to commit one to the very existence of the non-existing entity. After all, by the law of excluded middle the negation of the sentence must have the truth value TRUE.[5]

Russell's proposal is then to paraphrase sentences in which definite descriptions occur in classical first-order predicate logic in order to bring out the logical form of the sentences. Instead of the definite descriptions carrying the weight of determining the referent, this role is now taken over by bound variables. As a result, quantifiers do not themselves presuppose entities, and therefore assigning a truth value to a sentence (or the negation of a sentence) in which a definite description occurs does not commit to non-referring terms. Quine then argues that names, such as 'Walter Scott' or 'Pegasus' can be analyzed similarly when restated as descriptions. If it is not easy to come by a description, an unanalyzable one such as 'the thing that [artificial verb]' could be used. For instance, with a sentence in which an existence claim with regard to electrons is made, then 'the thing that [electronizes]' could be used as a description of the name a for the specific electron. When, say, a theory contains an existentially quantified sentence with a variable ranging over electrons, such as a, then it commits to electrons.

The criterion of ontological commitment is thus a formal criterion about what theories, or groups of theories, entail. This way, Quine steers away from relying on pre-theoretic opinions, or arguments based on what appears as evident to us by intuition (Hylton 2007, pp. 268–269). Relying on such, often non-rigorous use of, pre-theoretic opinions or intuitions can lead to all sorts of undesired consequences. So, to circumvent these with the suggested approach, this means that given the fact that the theories belonging to the natural sciences are considered to be true, then given the fact that they quantify over variables ranging over both physical and mathematical entities, this would mean that one would discover that one is committed to the mathematical entities quantified over just as well as to the physical entities quantified over, and that therefore science is platonistic.

Support for Thesis B is based on a comparison of the best mathematicized theories of natural science on the one hand and non-mathematicized theories within the same domain on the other hand, with the proviso that there are any of the latter to compare the former with. As mentioned in Sect. 1.3, it is not enough to observe that mathematics is used in the sciences. Rather, the claim is that there is no suitable way in which the scientific theories can be rewritten in a non-mathematicized way (Marcus 2014a, p. 346).

According to Putnam (1975), the entanglement of, for example, physics and mathematics is of such nature, that one cannot be a realist with regard to physics, and a nominalist with regard to mathematics at the same time. For instance, the functions relating physical terms to mathematical terms, such as we saw in the example of Newton's second law above, are integrated in the theory in such a way,

[5] As mentioned before, classical logic plays an important role here.

that it would make no sense to separate them and still be able to speak about, say, the measurement of the mass of a body (Putnam 1975, p. 74). Thus, the quantification over the mathematical entities is considered to be indispensable here:

> quantification over mathematical entities is indispensable for science, both formal and physical; therefore we should accept such quantification; but this commits us to accepting the existence of the mathematical entities in question (Putnam 1971, p. 57).

Quine also emphasizes the relation between physical theories and mathematics. Specifically, he stresses the fact that the mathematical entities contribute just as much as the theoretical entities in physical theories:

> Ordinary interpreted scientific discourse is as irredeemably committed to abstract objects—to nations, species, numbers, functions, sets—as it is to apples and other bodies. All these things figure as values of the variables in our overall system of the world. The numbers and functions contribute just as genuinely to physical theory as do hypothetical particles (Quine 1981a, pp. 149–150).

Taken together, Theses A and B lend support for the following first premise of the indispensability argument, as reconstructed by Colyvan (2001, p. 11), and whose formulation of the argument is often considered to be its canonical form:

P1 Mathematical entities are indispensable to our best scientific theories.

In terms of Quine's idea of quantifier commitment, there are two ways in which the above indispensability claim can be described, each leading to a different way in which the existentially quantified sentences occur in the formulas of the scientific theories. Baker (2003, p. 53) calls the first way the 'weak form', where the existential quantifier falls within the scope of the universal quantifier ($\forall\exists \ldots$), and the second way the 'strong form', where the universal quantifier falls within the scope of the existential quantifier ($\exists\forall \ldots$).

Depending on the formulation, one arrives at different conclusions. Accepting the weak form results in establishing that reference to (some) mathematical entities is indispensable in general. Without them, the best scientific theories could not be formulated. However, it fails to be conclusive about which specific entity or entities it commits to and hence, the only conclusion is that there are mathematical entities, full stop. Accepting the strong form on the other hand results in establishing the indispensability of some specific entity or entities, and therefore implying that the best scientific theories could not be formulated without those specific mathematical entities, and hence the conclusion is that there is *this* specific mathematical entity, or *these* specific mathematical entities.

The indispensability argument then continues with the second premise, namely:

P2 We ought to be ontologically committed to all and only those entities that are indispensable to our best scientific theories.

The "only those entities" direction of the above premise can be defended by appealing to the idea of *ontological naturalism*, which stems from the idea that philosophy is continuous with science. It is the

> recognition that it is within science itself, and not in some prior philosophy, that reality is to be identified and described (Quine 1981b, p. 21).

One of the key ideas is that all that exists is a spatio-temporal continuum, together with all the entities inhabiting it (Marcus 2014a, p. 3576). And the spatio-temporal realm is precisely the subject of study of the sciences. And as such, "ontological questions, under this view, are on a par with questions of natural science" (Quine 1951, p. 43).

Ontological naturalism is sometimes also understood in a negative way: accept only those entities in your ontology that the best scientific theories are ontologically committed to. This precludes, for instance, other entities that some metaphysical and/or theological theories are committed to.

The other, or "all ... entities", direction of the premise is supported by an appeal to *confirmational holism*, which is a relation between empirical evidence on the one hand and theories in their entirety on the other. It states that an individual (observational) statement, or individual hypothesis, cannot be tested in isolation. After all, an isolated hypothesis never entails an empirical prediction. The point is that in the background a whole theory, or group of theories, is at work, containing auxiliary hypotheses, laws, and other background assumptions. Deductive logic does not provide us with an answer as to which of those in the conjunction can be (dis)confirmed. Only the whole theoretical framework can be (dis)confirmed. This is what led Duhem (1906b, p. 260) to observe that an experiment "can never condemn an isolated hypothesis".

Given that it is a whole theoretical framework which is confronted with, and justified by, observational data, Quine (1960a, pp. 362–364) concludes that this evidence then also pertains to the mathematical part of the theoretical framework. Because considering the whole of a theoretical framework implies considering all of its individual components, and this therefore also includes the whole inferential apparatus, i.e. the logic and mathematics (Marcus 2014a, p. 3578). Importantly, the acceptance of confirmation holism also means that it is sufficient to establish the weaker claim where the existential quantifier falls within scope of the universal quantifier. The conclusion of the Quine-Putnam indispensability argument is indeed an endorsement of mathematical platonism:

C ∴ We ought to be ontologically committed to mathematical entities.

Once one accepts the premises, the conclusions follows. Central to the argument is the idea that certain properties can be transferred from a class of undisputed theories to a class of disputed theories:

> [s]uppose we have a class of theories of such a kind that we possess fairly clear means of establishing whether they are true (or justified, or confirmed, or the like) ... Suppose that we have theories of another disputed class, a class for which we lack those probatory means ... If we have a way of showing that appeal to theories of this second class cannot ... be avoided by theories of the first class, we may have a reason to think that some appropriate semantic or epistemic property ... is transferred from the former to the latter. This will give us indirect probatory means like those at our disposal for theories in the former class; this will then allow us to establish whether theories in the second class are true (or justified, or confirmed, or the like) (Panza & Sereni 2016, p. 470).

In the case of the Quine-Putnam indispensability argument, the specific goal is to establish whether the disputed objects of the one class of theories exist, by transferring some accepted properties of the undisputed objects in the other class of theories to the former class.

Although valid, the argument is naturally not without its objections. The kinds of objections to it can be roughly divided up in those who deny mathematics to be indispensable to our best scientific theories, *pace* the first premise; those who accept mathematics to be indispensable, but nonetheless deny ontological commitment to mathematical entities, *pace* the second premise; and those denying one direction of the second premise.

Of those who deny that mathematics is indispensable for science, the most prominent exponent is Field (1980). In his work 'Science without Numbers', he follows a suggestion by mathematician David Hilbert, and shows that large parts of Newtonian gravitational theory can be formulated without quantifying over mathematical entities. The brilliance of the work, although it only shows the dispensability of mathematics of a part of Newtonian physics, is that it also proves a representation theorem, which shows that none of the important notions pertaining to numerical claims in the original mathematical formulation of the theory, such as can be found in the functions for gravitational potential or mass-density, are lost in the meta-theory (Colyvan 2012, pp. 68–69). Of course, it needs to be seen that a similar strategy is possible for other parts of Newtonian physics, and, more importantly, for the more recent physical theories. Nonetheless, the result is far from trivial.

Given the monumental task of formulating scientific theories in a nominalistically acceptable manner, this course has been dubbed 'hard-road' nominalism by Colyvan (2010), as opposed to 'easy-road' nominalism, where the indispensability of mathematics is accepted, but whose adherents reject the claim that one needs to ontologically commit to mathematical entities. Notable proponents of the latter persuasion include Melia (2000), Azzouni (2004), and Leng (2010), although it must be said that their strategies differ.

Coming back to the different directions of the second premise of the Quine-Putnam indispensability argument, it is clear that the "all ... entities" direction, which is supported by confirmational holism, is the crucial one to derive the conclusion.

The reason why the "all ... entities" direction is crucial for deriving the conclusion results from the fact that when resorting to holism, the extension of evidence from the empirical to the mathematical is straightforward. Empirical evidence counts as evidence for the whole framework, and it therefore also counts as evidence for the mathematical part of it. As a consequence, whatever role the various posits play is, in a sense, immaterial. There is no prior sifting of such details.

But the role the different posits play does matter, argues Maddy (1992). For oftentimes the models make use of idealizations, or even plain falsehoods. In fluid dynamics, for instance, matter is assumed to be continuous; or in the analysis of water waves it is often assumed that water is infinitely deep (Maddy 1992, p. 281). Baker (2005, p. 224) echoes the point made by Maddy (1992), to wit the fact that we do *not* tend to accept in our ontologies entities posited by our best scientific models such as frictionless slopes, ideal gases and infinitely-deep fluids.

Take ideal gases. In Newtonian dynamics, these are used to predict the behaviour of gases especially at higher temperature and lower pressure. These ideal gases are composed of moving point particles whose only interactions are perfectly elastic collisions, which means that the total kinetic energy, which is the energy due to movement, stays the same for the colliding particles. In reality, this is not the case, but it allows scientists to abstract away from the conversion of kinetic energy to, for example, noise or heat.

Apart from the fact that one would not probably want to ontologically commit to such ideal posits, nor would one want to commit to what they imply, such as justifying the continuous or the infinite on the basis of the just mentioned, even perhaps indispensable, use of the mathematics in the respective theories. Of course, such criticism is not conclusive, because it could be argued that underlying these idealizations, there are other or more fundamental principles at work allowing us to see that the idealizations are indeed the idealizations they are. For Quine, this would for instance mean that idealizations can be paraphrased. And this paraphrase is ultimately what counts as the true description (Quine 1960b, p. 250). As Maddy (1997, p. 145) puts it: "a statement about the frictionless planes is paraphrased as a claim about the behaviour of actual planes as friction is gradually reduced to a minimum."

Nonetheless, this is not all there is to say about it, according to Maddy (1997). Because besides the use of idealizations in scientific theories, the question whether ontological conclusions can be drawn in all cases presses itself.

When theories are tested, scientific practice shows that not all posits are considered to be subject of the tests. So, not all posits are considered to be fit for drawing ontological conclusions because of the role they play in the scientific theories (Leng 2005, p. 176). For example, if one looks at the use of continuum mathematics, such as differential equations representing natural phenomena, then, adhering to the criterion of ontological commitment, the physical phenomenon must actually be continuous (Maddy 1997, p. 146). But this is not always obvious. Is spacetime to be regarded as continuous because continuum mathematics can be applied to it so readily? In other words, is it experimentally verified that spacetime is continuous, or is it merely useful, because it makes calculations easier (Maddy 1992, p. 282). It seems that room for pragmatic aspects and context dependent features needs to be made, so as to allow for a more finely meshed analysis of the different roles of the various components of the theories, be they physical or mathematical.

Another reason why platonists might have a preference for resorting to holism is to sidestep the issue of what it means for something to be indispensable (Baker 2005, pp. 223–224). Spelling out what indispensable means requires, after all, an analysis of the role of the different posits, again be they theoretical or mathematical. Given the considerations relating to idealizations, if one would want to give up holism, yet hold on to the indispensability argument, a criterion is needed to determine what it means for some posit to be indispensable. Quine's insistence on confirmation holism was after all, as alluded to before, also intended to avoid double talk, thereby proposing a way for ontological commitment that could in principle be acceptable by both nominalists and platonists (Marcus 2014b, p. 349).

But even if it can be spelled out in a principled way in what sense the different posits in scientific theories are indispensable, including the fact that the mathematics involved is indeed indispensable for the different scientific endeavours, one might still argue that this is, in general, not enough to allow for ontological conclusions to be drawn. Melia (2000, 2002) and Colyvan (2002) acknowledge the different roles the various posits can fulfill within the scientific theories, and agree that we only ought to be ontologically committed to those entities that are indispensable *in the right way*. According to them, echoing an earlier point made by Field (1989), being indispensable for scientific theories is not enough; they also need to be indispensable for scientific *explanations*. This rests on the idea that when theoretical virtues increase, resulting from the postulation of mathematical entities, then this is a good reason for accepting the theory. It is argued that especially explanation is one of those virtues providing a good reason to accept (at least part of) a theory, as providing explanations is one of the important roles of scientific theories. In any case, these kind of considerations signify a move away from confirmation holism.

2.2 The Enhanced Argument

The original indispensability argument (Quine 1948; Putnam 1971) is the backdrop against which Baker (2005) advances the EIA, in order to address the fact that the indispensability of mathematics, and specifically the indispensability of the mathematical entities being invoked, is in itself not sufficient for drawing ontological conclusions, as was discussed in Sect. 2.1. The original indispensability argument is thus 'enhanced' by restricting the set of entities under consideration for ontological commitment to those entities that are indispensable for scientific explanations.

The main difference with the original indispensability argument is that this version does not rely on confirmation holism, thereby intending to circumvent (some of the) difficulties the original argument faces. Instead, the enhanced version makes use of the principle of inference to the best explanation (IBE). A principle that is widely used in the philosophy of science (Molinini et al. 2016, p. 321). According to the principle of IBE, we have to accept the existence of those entities that feature in the best available scientific explanations of empirical phenomena. Van Fraassen (1980, p. 19) formulates the general idea as follows:

> Let us suppose that we have evidence E, and are considering several hypotheses, say H and H'. The rule then says that we should infer H rather than H' exactly if H is a better explanation of E than H' is.[6]

[6] Because the formulation is meant to capture the general idea of what IBE amounts to, Van Fraassen (1980) goes on to say that "[v]arious qualifications are necessary to avoid inconsistency: we should always try to move to the best over-all explanation of all available evidence."

According to Sellars (1963), inferring H over H' amounts to having good reasons for holding theory H, and

> to have good reason for espousing a theory is *ipso facto* to have good reason for saying that the entities postulated by the theory really exist (Sellars 1963, p. 97).

The aim of the EIA is thus to make use of the same kind of reasoning that is often employed in arguments for the existence of physical, or theoretical, entities, and to extend it to abstract entities.[7] The category of abstract entities includes, of course, mathematical entities. This way, the EIA intends to give empirical reasons for the existence of mathematical entities in a principled way (Busch & Morrison 2016, p. 436).

It is clear that this move is directed at those who accept the principle of IBE in the first place. Predominantly, those who can be categorized as the so-called 'scientific realists' do accept the principle of IBE (Hunt 2016, p. 453).[8] Scientific realism is the position that scientific theories, in mature sciences, are considered to be approximately true, or at least aim at truth, and that as a consequence the fact that one accepts a scientific theory implies that one believes the theory is true (Van Fraassen 1980, p. 12).

Therefore, scientific realists should uphold, according to proponents of the EIA, that the same inferential evidence which is used in the best explanations of empirical phenomena, specifically for accepting the existence of theoretical entities, should also count as inferential evidence for accepting the existence of the abstract entities in the theory, on pain of being accused of what Putnam (1971) calls 'intellectual dishonesty'.

Thus, the idea of the EIA is that if there are mathematical entities that fulfill the same kind of explanatory role as theoretical entities in scientific theories, and if commitment to theoretical entities is defended by means of the principle of IBE, then this very same inference should also apply to all the entities referenced in a scientific theory. And therefore, mathematical entities are on par with theoretical entities in these cases (Saatsi 2011, p. 144).

The EIA can probably be best be summed up in its more or less canonical form (Baker 2009, p. 613):

P1 Mathematical entities play an indispensable explanatory role in science.
P2 We ought to be ontologically committed to any entity that plays an indispensable explanatory role in our best scientific theories.
C ∴ We ought to be ontologically committed to mathematical entities.

The enhanced argument is indeed stronger than the original indispensability argument. In the same explanations in which theoretical entities are postulated, and in which the theoretical entities are essential for the explanation, there are often

[7] The fact that this kind of reasoning is used in arguments for the existence of theoretical entities does not imply that it is the only way in which ontological commitment is determined.

[8] As to be expected, there are various issues with the principle of IBE. Some of these will be discussed in Sect. 3.4.

also mathematical entities postulated, which in turn are then also essential for the explanation, and therefore "if inference to the best explanation licenses one it licenses the other" (Field 1989, p. 17).

Depending on how exactly the argument is viewed, it can be seen as an extra opportunity for the nominalist or as an extra opportunity for the platonist (Marcus 2014a, p. 351). In case of the former, even if it is not possible to have non-mathematicized formulations of scientific theories, this observation by itself does not establish platonism anymore. Apart form showing that the mathematics is indispensable for the theory, it also has to be shown that the mathematics is indispensable for the explanation of the phenomenon in question. In case of the latter, even if it is possible to have non-mathematicized formulations of scientific theories, this in itself does not establish nominalism either. It also needs to be shown that the mathematics is dispensable for the explanation.

In any case, the possibility for a successful EIA stands or falls with the possibility of finding a case of a genuine extra-mathematical explanation (Baron 2020, p. 19). So, the challenge is to find an example of such a mathematical explanation of a physical phenomenon.

2.3 External Application

Before continuing to the discussion of an example of an extra-mathematical explanation, let us first have a look at some of the potential difficulties that putative examples face.

One of the reasons for the focus on the application of mathematics in science in relation to the EIA is that if an IBE is to be of any use at all, the inference needs to be assumed true. Otherwise, there is no phenomenon to be explained.[9] If the EIA is indeed meant to be an argument for the existence of mathematical entities, then this in turn means that, in order to prevent the argument from becoming circular, the explanandum must itself be free of any reference to mathematical entities, stemming from the fact that it must be assumed true (Baker 2005, p. 225). Intra-mathematical explanations, such as in explanatory proofs, are therefore excluded. Their explananda all contain references to mathematical entities. For this reason, external applications of mathematics must be used. It should, at least in principle, be possible to find examples of explanations that do not have references to mathematical entities in their explananda. Consequently, a detour needs to be made through the sciences.

When looking for examples of mathematical explanations of physical phenomena, the first science that probably comes to mind is physics. After all, the application of very sophisticated mathematics in physics abounds, and indeed,

[9] Notably, although in a different context, Van Fraassen (1980) and Cartwright (1983) object to the idea that explanations necessarily have a true explanans.

the mathematics may well be indispensable. However, there are some difficulties with using examples from physics when in comes to explanatory indispensability arguments.

One first observation is that the nature of physics is highly theoretical. As a result, one of the issues is that it can be difficult to determine the boundary between the physical and the mathematical in examples taken from physics:

> The divide between the mathematical and physical components of such examples is often unclear [and] ... [i]f the distinction between abstract and concrete is controversial, or open to interpretation ... then the force of the examples is thereby undermined (Baker 2015, p. 243)

Similarly, when Maddy (1992, p. 282) raised the question whether the continuous character of spacetime is experimentally verified or whether it is the result of the use of continuum mathematics, because it is so readily applicable, the question is where to draw the line between the mathematical and the physical. In a similar vein, Steiner (1978b, p. 19) asks us to "[i]magine defining rate of change without the resources of analysis." Price (2017, pp. 40–42) also emphasizes the oftentimes intimate relation between a theoretical framework and the facts in physics. Specifically, in certain cases it appears that it is only possible to identify a fact given a background theory, and that, as such, the explanandum is not independent of the explanans. Indeed, this echoes the just mentioned point made by Maddy.

The difficulty of drawing a clear and unequivocal line between the mathematical and physical is problematic for an argument that is meant to establish the existence of mathematical entities via IBE. Not only is the force of the examples undermined, as Baker observes, but in the worst case the argument will become circular and as a result be of no use.

A different worry for using examples from physics pertains to the status of geometry. Take for instance the explanation of the bending of the path of light when close to an object of great mass. The explanation for this phenomenon can be found in general relativity, which states that light, when moving freely, follows a geodesic in curved spacetime. The path of light is thus a result of the curvature of spacetime. But are we talking physical geometry, or axiomatic geometry? In other words, is the geometric explanation mathematical or concrete? If it turns out that the explanation can do without pure mathematics, i.e. axiomatic geometry, then it can be argued that the explanation is not genuinely mathematical (Colyvan 2001; Baker 2005; Marcus 2014a). Similarly, the explanation of the Lorentz contraction of bodies in motion relative to an inertial reference frame depends on geometry for its explanation (Colyvan 2001).

Of course, this is not to say that there could not be sound examples from physics pertaining to geometry. Or that there could be examples in which the distinction between the mathematical and physical in the model is clear in the relevant sense. Nonetheless, it is precisely the highly mathematical nature of the field that complicates the separation of the concrete and the abstract to a certain degree.

Another issue, briefly touched upon earlier, that limits the class of putative examples is the usage of idealization in explanations of physical phenomena. For example, the explanation[10] of the existence of two antipodal points on the earth's surface having the same temperature and pressure at any given time t (Colyvan 2001, p. 49). Algebraic topology, due to a corollary of the Borsuk-Ulam theorem, explains why this is so: for every continuous function $f : S^n \to \mathbb{R}^n$, there is a mapping of a pair of antipodal points to a point at the surface of a sphere. The earth is modeled as a topological sphere of the form S^2 (and temperature and pressure are treated as continuous functions), with point particles mapped to the real numbers (Leng 2005, p. 181). The model of the earth clearly is an idealization, meant to pick out some relevant features of the earth. This is in itself not problematic, but in the EIA the explanans is also assumed to be true, and the idealization does not find its counterpart in the physical world.

Cartwright (1980, p. 75) goes even on to say that laws of physics are predominantly false. Her point is that the goal of explanation is different from the goal of accurate description, and that these goals are not always reconcilable. One example she gives is Newton's law of gravitation, which states that the force between two bodies varies inversely as the square of the distance between them, and is proportional to their masses. But this law can only deal with very simple, or ideal(ized) cases. In cases where a body is charged, for example, then Coulomb's law, which states that the force between two charged bodies varies inversely as the square of their distance and is proportional to their charge, is also needed to determine the total force (Cartwright 1983, pp. 57–58).

This is not to say that a law cannot describe the facts. In cases where (it is assumed that) there is one, say, force at work, then the law can be said to be factive. But if the laws are considered to be factive in cases where multiple forces are at work, then the problem is that in those cases the laws must actually take the form of a conditional statement. Cartwright gives the further example that, if the bodies are not charged etc. ..., then $G = Gm_1m_2/r^2$. But then the law's relevance is severely limited, and it does not explain the actual situation. The trade-off is, she concludes, that in cases where the laws are true, they explain relatively little, and in cases where they do explain, they oftentimes do not state the facts (Cartwright 1983, p. 72). Accurate description on the one hand and explanation on the other oftentimes appear to part ways.

The tenacity of such and other difficulties marked a shift towards biology. As a field, biology is much less mathematicized than physics, but where mathematics is still used, for instance for the formulation of models for optimization strategies. The idea is that it therefore will perhaps lend itself better for finding a case of a genuine extra-mathematical explanation.

[10] Leng (2005) argues that the example would constitute a prediction rather than an explanation, because, strictly speaking, no such two points have been discovered, and due to the extreme level of detail needed to do so, this is unlikely to happen.

2.4 An Extra-Mathematical Explanation

In his (2005) paper, Baker presents a case in which he claims to have found an example, in evolutionary biology, of a mathematical explanation of a physical phenomenon. The example is supposed to push the debate forward, because it is believed that it does not suffer from the difficulties such as the ones mentioned in the previous section.

The example treats of the Periodical Cicada in North America, which is a genus of the Cicada family, of which the species have either a 13- or a 17-year life cycle. During this life cycle, they spend all but two weeks underground in larval form. During the last two weeks of their lives they collectively come above ground, in their adult form, for the sole purpose to reproduce. Biologists are interested in these odd periodical life cycles and try to find explanations for them. It might for example be the case that prime-numbered life cycles would give them some evolutionary advantage. Biologists tend to think that the reference to the primality of the life cycles plays a pivotal role:

> [p]eriodical cicadas present numerous puzzles for biologists ... [T]he 13- or 17-year periods of cicadas suggest there is something important about prime numbers (Lehman-Ziebarth et al. 2005, p. 3200).

And more specifically, biologists use prime-number talk in (possible) explanations of the occurrence of the observed length of the life cycles:

> the persistence of prime-numbered cycles has been shown to be a plausible outcome of circumstances in which already-periodical cicadas face disproportionately high predation pressures when they emerge at low densities. Moreover, repeated instances of life-cycle switching from 13 to 17 years or vice versa over evolutionary time suggest that these life cycles are stable strategies, whereas other alternatives are not (Ito et al. 2015, p. 2).

Thus, the thing to be explained, the explanandum, is why the life cycles of these species are prime. Baker gives the basic structure of the explanation (Baker 2005, p. 233):

(1) **Biological 'law':**
Having a life cycle period which minimizes intersection with other periods is evolutionary advantageous.
(2) **Number-theoretic theorem:**
Prime periods minimize intersection (compared to non-prime periods).
(3) **'Mixed' biological/mathematical law:**
Hence organisms with periodic life cycles are likely to evolve periods that are prime.
(4) **Ecological constraint:**
Cicadas in ecosystem-type, E, are limited by biological constraints to periods from 14 to 18 years.
(5) **Species specific prediction:**
Hence cicadas in ecosystem-type, E, are likely to evolve 17-year periods.

In the context of indispensability arguments in general, and the EIA in particular, the step in the argument that is of specific interest to us is step (2).[11] As Field (1989, p. 15) puts it:

> one of the assumptions that appears in this explanation is claim S, and we are pretty sure that no explanation of the phenomena that does without claim S is possible.

In the case at hand, the reference to the number-theoretic theorem in step (2) is claim S. Furthermore, it is an 'enhanced' indispensability argument, according to Baker (2005, p. 237), because the number-theoretic theorem (2) has a genuine explanatory role in relation to the periodicity of the life cycles, in a way quite different from, for example, idealized concrete objects.

2.5 Conclusion

At the start of the chapter, objections to the idea of confirmation holism in the original Quine-Putnam indispensability argument were discussed in Sect. 2.1. Given that different posits do actually play different roles in scientific theories, confirmation holism needed to be given up. This raised the question of what being indispensable could then mean, and if indispensability was enough for ontological conclusions to be drawn. The conclusion was that indispensability was not sufficient in and of itself, but that the entities concerned should be indispensable in the right way, namely indispensable for explanations.

This opened the door for an explanatory version of the indispensability argument, which was presented in Sect. 2.2, and which should push the debate forward. After having discussed some difficulties potential candidates face in Sect. 2.3, a concrete example from biology was presented in Sect. 2.4. This example appears promising, as it does not appear to suffer from many of the difficulties from which earlier candidates suffered. However, as we shall see in the subsequent Chaps. 3 and 4, other objections can be raised.

[11] In (2017a, pp. 783–784), Baker gives an alternative version of the argument. But the first three steps are the same as in the initial formulation.

Chapter 3
Objections

Abstract Before continuing with the evaluation of the internal logical structure of the enhanced indispensability argument, this chapter discusses objections that are relevant for arguments discussed in subsequent chapters, such as explanatory generality, compatible explanations, representation and inference to the best explanation.

The EIA has generated a lot of discussion,—see, for instance, (Molinini et al. 2016) for a good overview. Before continuing with the evaluation of the internal logical structure of the argument, there are some objections that need to be mentioned beforehand. There are of course many more objections, but to give a complete analysis of the current state of the discussion, and to list all the objections to the argument, would be a project of a completely different kind and surpass the immediate topic we are concerned with. However, there are four objections in the literature that are relevant to include here. The first two of those relate directly to some of the arguments given in subsequent chapters. Section 3.1.1 presents a criticism directed at the often made claim by proponents of the EIA that mathematical explanations are better than their nominalistic counterparts because of their generality. Section 3.2 discusses the possibility of compatible explanations and the problem this poses for the original indispensability argument and by extension its enhanced version. If different mathematical formulations of scientific explanations are possible, then the EIA cannot fulfill its ontological promises. The section also briefly traces the historic accounts where the present objection took inspiration from.

The second two objections are of a more general kind, but nevertheless warrant a brief survey. Section 3.3 discusses the argument that mathematics is not explanatory, but rather, for instance, representational. What carries the explanatory load are the physical facts, or so the argumentation goes. Section 3.4 will briefly mention some objections to the principle of IBE, since the EIA makes essential use of it. Section 3.5 summarizes the chapter and states the important points in relation to the subsequent chapters.

© The Author(s), under exclusive license to Springer Nature Switzerland AG 2025
L. A. Tump, *On Mathematical Explanations of Empirical Phenomena*,
Synthese Library 517, https://doi.org/10.1007/978-3-031-98860-8_3

3.1 Explanatory Generality

Among the often appealed to virtues in defence of the explanatory superiority of mathematical explanations in science to their nominalistic counterparts, often the greater explanatory generality of the former compared to the latter is mentioned. The mathematical explanations are considered to be better explanations, because the mathematics used makes them more general, by allowing for the unification of different phenomena by either topic or scope generality (see Sect. 1.3).

One assumption behind the appeal to generality is that the difference in explanatoriness ultimately comes down to the fact that the mathematics represents more general features, or more general explanatory features, of reality. The nominalist alternatives on the other hand, it is then argued, lack these unifying abilities and, as a result, are not able to represent these more general features of reality. The link between some form of unification and explanatory strength has been emphasized over the years by, for instance, Baker (2017b, p. 199) and Colyvan (2002, p. 72) in relation to the current debate, but also independently of it by proponents of unificationist theories of explanation, such as Kitcher (1989).

Knowles and Saatsi (2019, p. 1135), on the other had, observe that an explanation can become more explanatory without accurately representing further explanatory features of reality, and argue that topic generality is not explanatory at all, and scope generality only up to a point (Knowles & Saatsi 2019, p. 1122).

3.1.1 *The Limits of Generality*

Beginning with scope generality, Knowles and Saatsi (2019) argue that the fact that the scope of mathematics is greater (and even without limit) than the nominalistic counterparts, this in itself is no reason for judging a mathematical explanation to be better on those grounds alone.

To begin with, they observe that although mathematics has the ability to deal with infinitely many cases, most of the cases are too far removed from the scientific or theoretical context of the explananda (Knowles & Saatsi 2019, p. 1128). In terms of the dialectical situation of the EIA, this is an important observation, because the argument operates from within a broadly naturalistic framework.[1] Therefore, only (explanatory) cases relevant to either the empirical evidence or the context of the scientific theory in question should be considered. If an explanation makes use of, say, number theory, then it is clear that it is limitless in the sense that the explanation can make use of any number, no matter how large. But, they argue, most natural numbers exceed the empirically relevant context and cannot be evaluated outside a mathematical context. In case of the life-cycles of the Periodical Cicada,

[1] A similar observation will be made in Sect. 4.2.

for instance, the naturalistic framework of evolutionary biology is not sufficient for evaluating such extreme cases (Knowles & Saatsi 2019, p. 1128).

To motivate their argument from a more neutral point of view, Knowles and Saatsi (2019) invoke the counterfactual theory of explanation (cf. 1.1). The counterfactual theory of explanation is motivated independently of the debate of the EIA, and would therefore not be question begging for that reason. Also, the counterfactual theory of explanation can be made to fit the naturalistic context of the debate on the EIA (Knowles & Saatsi 2019, p. 1122). Within said theory, scope generality would then correspond to the range the variables of the explanans can have. Besides, an explanatory relationship over the range of these variables needs to hold (Knowles & Saatsi 2019, p. 1123).

To put it in other words, it is the range of 'what-if-things-had-been-different' questions that can be asked in relation to some phenomenon that is to be explained. However, if the (limitless) scope of the mathematics invoked is used as an argument for the explanation being better than a nominalistically acceptable one, i.e. the larger scope of the former, then Knowles and Saatsi (2019, p. 1127) are quick to point out that evaluating such extreme counterfactuals cannot be done without some modal assumptions operating in the background that would surpass the naturalistic context. For instance, the earth could have been somewhat different, e.g. smaller or larger in size, which in turn could have affected, say, the length of the life cycles. But such a life cycle would still be finite. If cycles larger than, say, the age of the universe are concerned, let alone infinitely long life cycles, then a theory needs to be in place to be able to accommodate those. But this, they argue, is far removed from the actual explanandum and the biological context it is part of. In fact, it is not clear at all how these extreme counterfactuals can contribute to the explanation of the phenomenon in question, given the context in which the question was initially asked. Thus, although an increase in scope generality can result in a better explanation to a certain degree, this, given the naturalistic context, is not something that a nominalistic alternative cannot in principle achieve.[2]

Topic generality, in the context of the counterfactual theory of explanation, corresponds to the possibility of asking 'what-if-things-had-been-different' questions about different subject matters. The explanation thus connects disparate subject matters (Knowles & Saatsi 2019, p. 1123). Where scope generality is able to improve an explanation to a degree, it is not immediately clear how topic generality would make an explanation better.

If, say, two disparate phenomena that where explained separately before can be subsumed under a common explanation, then the question is how this would make this new explanation better? It might be argued that the explanation is better, precisely because it *unifies* two previously unconnected topics. But if this is so, Knowles and Saatsi (2019, p. 1129) argue, then this is down to the fact that the topic generality resulted in an increase of scope generality, that is, it resulted in an increase in the range of explanans variable values for each of the phenomena individually.

[2] For an example of a nominalistically acceptable alternative, see Sect. 4.2.

Because if it is not for the increase in scope generality, then, given the counterfactual theory of explanation, the number of (potential) 'what-if-things-had-been-different' questions of the one phenomenon did not increase because the 'what-if-things-had-been-different' questions of the other phenomenon which in the new explanation is structurally related to it. Besides, an increase in topic generality could also result in lesser number of counterfactual questions, if topic specific details are lost in the more topic general explanation. After all, topic specific explanations are very well suited for the domain specific phenomenon under investigation (Knowles & Saatsi 2019, p. 1129).

Even though explanations can exhibit topic or scope generality, this in itself does not justify the claim that this increases the amount of ontologically committing information compared to nominalist alternatives (Knowles & Saatsi 2019, p. 1128). Although scope generality can indeed increase the strength of an explanation, given the naturalistic context of the EIA, this is not beyond what is in principle achievable by a nominalistically acceptable alternative explanation. After all, how much modal information is relevant depends in part on the theoretical context. An explanation containing limitless modal information exceeds what could be reasonably asked from within a naturalist framework, and hence beg the question against the nominalist.

It might be argued that, although scope general or topic general explanations might not be better in the sense argued above, they do pick out more general facts of reality. Specifically, they pick out general explanatory features, or certain dependencies, which lead to a deeper understanding of the phenomenon in question. Each individual fact is as it were embedded in this more general fact, i.e. the fact or facts of the more general and unifying mathematical explanation. Thus, the explanation of the individual facts holds because of the overarching more general fact. A nominalistic alternative would not be able to explain why the individual facts hold, or why a certain more general pattern emerges. But on the counterfactual analysis of explanation, the individual facts are not *dis*unified. The counterfactual dependence brings out their relationship (Knowles & Saatsi 2019, p. 1129). As an aside, the same goes for the nominalistic account of for example Field (1980) or for the account presented in Sect. 4.2.

In any case, any argument for the claim that mathematical explanations are better than their nominalistic counterparts may not start from the assumption that they are better because the mathematics provides more explanatory depth per se. Specifically, those who defend the idea that explanatory patterns hold because of some mathematical fact need to take into account the naturalistic context within which the EIA operates.[3]

[3] Baron, Colyvan, and Ripley (2017) develop an account of a counterfactual analysis of mathematical explanation which is supposed to bring out the counterfactual contribution of the mathematics as such. The counterfactual is effectively a counterpossible in their account. In that sense, they intend to unify intra- and extra-mathematical explanation. Knowles and Saatsi (2019, p. 1129) point out that if such an account is able to bring out the more general explanatory relationships, it needs to be shown that these explanatory relationships are missing in a (non-counterpossible) counterfactual

3.2 Compatible Explanations

Another issue the EIA faces is the problem of compatible explanations (Hunt 2016). The problem of compatible explanaitons gets its inspiration from the problem of multiple foundations for the traditional indispensability argument (Baker 2003), which in turn gets its inspiration from the problem of multiple reducibility of numbers as stated by Benacerraf (1965) in his famous work 'What numbers could not be'.

Hunt (2016, pp. 460–461) first observes that because of the use of an IBE, the EIA has to establish the explanatory necessity of some mathematical entity, not mathematics in general. He then observes that, given the use of an IBE in the EIA, mathematical entities do not compete in the same way as theoretical entities, or physical unobservables, do. In fact, there seems to be a lack of competition (Hunt 2016, p. 462). He then concludes that because of this lack of competition, and the possibility of multiple mathematical formulations of the mathematical entities referred to in science, an IBE cannot establish that some mathematical entity is indispensable for explanation. He also adds that it cannot refute nominalism either (Hunt 2016, p. 463).

3.2.1 Multiple Reducibility and Foundations

Let us first, for historical reasons, very briefly turn to the problem of the multiple reducibility of numbers. When viewed as sets, and given some suitable successor function, natural numbers can be reduced to different ω-sequences (Benacerraf 1965, p. 55):

(i) $\{\varnothing\}, \{\varnothing, \{\varnothing\}\}, \{\varnothing, \{\varnothing\}, \{\varnothing, \{\varnothing\}\}\} \ldots$
(ii) $\{\varnothing\}, \{\{\varnothing\}\}, \{\{\{\varnothing\}\}\} \ldots$

The observation by Benacerraf is that both reductions of the numbers are equally correct, in the sense that they can both prove the same theorems.[4] This means that numbers can be reduced to sets in multiple ways. Hence, the platonist's attempt to reduce numbers to sets to bring about their one true nature is inherently problematic.[5]

account. Also, the account must stay within a nominalistically acceptable framework, vis-à-vis the appealed to modalities. Chapter 7 will discuss a different problem an account that makes use of counterpossibles faces.

[4] The first ω-sequence (i) is known as the Von Neumann ordinals, and the second (ii) as the Zermelo ordinals.

[5] There is also the question as to whether numbers are sets, or whether the mentioned ω-sequences are about a representation theorem, i.e. that both structures are isomorphic. If one thinks more in terms of representation theorems, and not in terms of sets, than the argument presented here loses strength. More generally, depending on the position one holds in the philosophy of

In relation to the traditional indispensability argument, Baker (2003, pp. 55–56) argues that although it might well be that numbers are dispensable because they can be reduced to sets in multiple ways, the sets themselves would not be dispensable. Peano arithmetic, for instance, will in either case still be reducible to set theory. The indispensability argument would therefore still proceed as intended, but is then about sets rather than numbers.

Whatever the implications for the ontological status of numbers, what Baker takes away from this for the indispensability argument is not that multiple reducibility in itself poses a problem for it, but rather that the availability of alternatives to set theory exist that could act as foundations for arithmetic (Baker 2003, p. 57). And the problem with the existence of alternatives to standard Zermelo-Fraenkel (ZF) set theory for the foundations of arithmetic is, as Baker sees it, that in that case the indispensability argument cannot uniquely determine the indispensability of sets.[6] Baker (2003, p. 57) states that category theory is an example of an alternative foundation of arithmetic,[7] and that the ontology of category theory does not overlap with that of set theory. For this reason he concludes that there is no way in which to decide what foundation is to be preferred over the other, and that therefore all we have is a disjunctive commitment: we are committed to either sets or categories (Baker 2003, p. 58).

This is where Hunt (2016) picks up the debate in relation to the EIA. Even if it is deemed controversial that category theory (or set theory for that matter) could function as a foundation for arithmetic, what is important is that set theory and category theory provide different, non-overlapping, formulations of the application of mathematical entities in scientific explanations (Hunt 2016, p. 463).

3.2.2 Non-overlapping Formulations

As alluded to in Sect. 2.1, the scope of the indispensability argument can be formulated in two different ways, namely in a 'strong form' or in a 'weak form'. In the strong form, the existential quantifier has scope over the universal quantifier (i.e. being of the form $\exists\forall \ldots$). This means that there needs to exists a specific mathematical entity (or multiple mathematical entities) that plays a crucial role in all explanations of some phenomenon. In the weak form, the universal quantifier has scope over the existential quantifier (i.e. being of the form $\forall\exists \ldots$). Contrary to the

mathematics, such question will be answered differently. For example, a structuralist might argue that underlying the different representations is the same structure, and that therefore there is no multiple reducibility.

[6] As an aside, there are also alternatives within set theory. For instance Quine's New Foundations. The question is whether they compete with each other or whether they are equiconsistent, i.e. that the consistency of the one implies the consistency of the other.

[7] The question as to whether arithmetic needs a (set or category theoretical) foundation or not is a topic in its own right.

strong form, the weak form does not require reference to a specific mathematical entity. It just requires that every explanation references some mathematical entity (Baker 2009, p. 616).

Hunt (2016) argues that the availability of compatible explanations not only prevents an IBE from establishing the indispensability of a particular entity, i.e the strong form, but also from establishing a refutation of nominalism, i.e establishing the weak form. He states the problem of compatible explanations as follows (Hunt 2016, p. 461):

(1) To refute nominalism, the EIA must establish that a particular mathematical entity is indispensable to all adequate explanations of a given phenomenon.
(2) Inference to the best explanation cannot establish the indispensability of a mathematical entity if a compatible explanation does not reference that entity.
(3) Set theory and category theory provide alternative formulations for all mathematical entities applied in science.
(4) These alternative formulations provide compatible explanations for any scientific explanation that references mathematical entities.
(5) ∴ Inference to the best explanation cannot justify the indispensability of any mathematical entity applied in science. Hence, the EIA does not refute nominalism.

Premises (1)–(4) in Hunt's argument state that platonsim cannot be established by the EIA, based on the requirement that a specific mathematical entity must be indispensable in the explanation of some phenomenon in order to justify its existence, as opposed to the reference to different mathematical entities in all alternative formulations of the available adequate explanations. The justification for the requirement in (1) is that scientific realists use IBE to argue for the existence of theoretical posits only if these are ineliminable from explanations. Although eliminability does not equate indispensability, it is nonetheless so that if an entity can be eliminated by, say, referencing another entity instead, then certainly the that entity is not indispensable in the explanation. If the IBE that is used to argue for the existence of theoretical posits is to be used in the same way as in the argument for the existence of mathematical entities, then those that defend the EIA need to establish the strong form of the argument (Hunt 2016, p. 461).

But if multiple alternatives are available, as stated by premise (3), then it has to be decided via an IBE which of the alternatives is best (or perhaps more likely to be the best). But with regard to mathematical entities, Hunt (2016, p. 462) argues that there is not the same kind of competition, or even any competition at all, as is the case with theoretical entities: mathematical entities do not causally interact with each other, as opposed to physical entities. And as such, there is no prima facie problem for them to coexist. Sklar (1974), his book Space, Time, and Spacetime, already observed the possibility of there being multiple consistent mathematical descriptions of spacetime possible, which can resullt in descriptions of the world with very different geometric structures (Sklar 1974, p. 2). The situation for theoretical entities to coexist is different, because the competition between those is such that they exclude the possibility for the other entity to exist at the same time, given the spatio-temporal continuum.

Due to the lack of this kind of competition between mathematical entities, theoretical virtues are not in a position to adjudicate between any competing formulations. Even if one alternative formulation fares better with regard to some theoretical values than another, the lack of competition prevents these values to play a role in terms of ontological difference makers. This echoes a point made by Lipton (2004, p. 63), when he states that in the case of compatible explanations, IBE does not prevent allowing for inferring all of these compatible explanations at once.

Returning to the weaker "$\forall\exists$ ...claim", Hunt (2016, pp. 459–461) proceeds to argue, *pace* (Baker 2009), that the fact that there are compatible formulations that all make use of mathematics does not refute nominalism either. Contrary to the traditional indispensability argument, which accepts confirmation holism and thereby puts all entities within the scope of the quantifier epistemically on par, the EIA cannot do so. To be able to apply IBE in the context of the EIA, the entity in question must play an indispensable explanatory role. Accepting multiple compatible explanations, even if IBE in itself does not prevent one from doing so, does not show that any of the mutually substitutable entities is indispensable for the explanation. For this reason, he concludes, IBE is prevented to fulfill its intended role in the EIA (Hunt 2016, p. 459).

3.3 Mathematics Is not Explanatory

One of the central tenets of the EIA is that mathematics fulfills a genuine explanatory role with regard to (at least some) empirical phenomena. As mentioned in Chap. 2, the role mathematics plays not only needs to be shown to be indispensable, but it also needs to be shown to play this genuinely explanatory role (Daly & Langford 2009, pp. 641–642). Some argue that mathematics is not explanatory in itself (Melia 2000; Daly & Langford 2009; Saatsi 2011). Rather, they are the physical facts themselves that carry the explanatory weight. Mathematics rather *indexes* or *represents* the facts. And whether mathematics does or does not explain, what could its contribution to the formulation of explanations then be? Knowles and Saatsi (2019) suggest that the efficiency with which mathematics can present information increases our expressive capacity.

3.3.1 Mathematics Is Representational

Let us give the general idea of the representational role of mathematics by means of a couple of simple examples (keeping in mind the examples do not necessarily generalize). Take the measurement of the distance between two objects. Melia (2000, pp. 472–473) states that when the result of a measurement is expressed (as in '*a* is at a distance of 3 meters from *b*'), the mathematics, in casu the use of the number '3', merely indexes the spatial relationship between the two objects, and is itself not part of the relationship proper. Similarly, Balaguer (1998, p. 133) argues that, when a certain temperature of some physical system is measured (as in 'the physical system *S* is forty degrees Celsius'), the mathematics used in such a case, for instance with the invocation of the number 40, is not part of the physical system itself either.

With regard to measurement, things may often well be as straightforward as they seem here. And, in fact, there probably isn't any problem with that in many cases. After all, often there is indeed some arbitrariness in the choice of which unit of measurement to use. What remains in the case of the measurement of the distance between objects is the spatial relation, and in the case of the measurement of the temperature of the physical system the actual temperature of the system.[8] Moving up the complexity ladder, this arbitrariness becomes less obvious and, some argue, at some point altogether disappears. Baker and Colyvan (2011), for instance, argue that in more complex explanations, such as the Borsuk-Ulam theorem mentioned in Sect. 2.3, the mathematics cannot be dispensed of, because it is an integral part of the explanation.

But even in some cases involving measurement, such as the measuring of the life cycles of the Periodical Cicada in years (cf. Sect. 2.4), the argument for arbitrariness seems hard to make. It appears, at least from the perspective of biology, that there is something relevant to these specific cycles, that is, the life cycles in years as expressed by the numbers.

With regard to the explanatory role of mathematics in such cases, Saatsi (2011, p. 144) argues that the point is not so much that mathematics cannot play a role in our understanding of the world, but that there often is no clear distinction made between *explanation* and *representation*. By representing, mathematics can help us to get insight and understanding. However, this in itself does not imply that mathematics goes beyond its representational role, that is, that mathematics is explanatory of the phenomenon that it represents. Although the distinction between representation and explanation appears to be widely recognized, the problem is that the physical interpretations of the mathematical formalisms are not always separated properly from the mathematical formalisms themselves (Bueno 2012, pp. 967–969). This can lead to the idea that the mathematics itself does (some of) the explaining. Take the following argument by Bueno and French (2012, p. 104):

> Consider the example of a stone thrown into the air. At one point in time, the mathematical equation that describes the stone's movement has value zero. Does the fact that the equation has such a value provide an explanation of why the stone is at rest, or does it simply offer a mathematical description of the phenomenon in question? Presumably, no one would consider the fact that an equation has value zero to be by itself an explanation of a physical phenomenon. A suitable physical interpretation, which identifies the relevant physical processes responsible for the production of the phenomena in question, is needed in order to yield a satisfactory explanation.

[8] It must be noted that the notion of 'temperature' is also already quite a complex one.

On this view, the mathematics is a description of the phenomenon rather than an explanation of it.

What is essentially suggested here, *pace* Quine's criterion of ontological commitment, is to separate quantifier commitment from ontological commitment. Because of the representational role of mathematics, even if it were the case that mathematics plays an indispensable role (in explanations), it does not follow that one is committed to the mathematics invoked. The idea of separating quantifier commitment from ontological commitment can typically be found in so-called 'fictionalist' positions. Fictionalism, in the philosphy of mathematics, states that mathematical statements are not literally true, even though they are statements about mathematical objects, because abstract objects do not exist (Balaguer 2009, p. 132). This does however raise the question what the criteria for ontological commitment would be in this case, and whether a position such as this leads to what Quine called 'double talk' (see Sect. 2.1).

Regarding the alleged double talk, Bueno (2012, p. 976) argues that this need not necessarily be the case. He observes that in interpretations in the tradition of the likes of Frege and Quine, the existential quantifier does not distinguish between quantification within a domain, and the existence of some object in that domain. That is to say, no distinction is made between the 'there is an x' (or 'for some x') and the 'there exists an x' when quantifying over objects in a domain. And indeed, the lack of this distinction reminds us of Quine's conclusion when he stated, in the previously mentioned slogan, that "to be is to be a value of a bound variable". If on the other hand, as is for instance done in logics that deal with non-existent objects, a distinction between those two aspects is made, for example by introducing an existence operator (E) for predicating(!) existence, then the traditional existential quantifier ($\exists$) does not need to carry any ontological import (Bueno 2016, p. 2600). This lack of ontological import will then also concern those statements in which one quantifies over mathematical objects.

Having something like the above in mind, Bueno (2012, pp. 980–981) then emphasizes the fact that a suitable interpretation needs to be given to relate the mathematical to the physical. This interpretation needs to clarify the precise nature of the relation. Ultimately though, it is argued, the explanation itself rests on the physical phenomenon. On this view, mathematics is used to represent the physical phenomenon, mediated by a suitable interpretation. The status of this relation should at least in part account for the 'unreasonable effectiveness'[9] of the use of mathematics and its models in the natural sciences. Such an account can indeed not be left hanging (Bueno & Colyvan 2011, p. 351).

[9] See Wigner (1960) for the famous historical article on this issue. Many responses have been given to it over the years, and not everyone agrees that the effectiveness of mathematics in the natural sciences is quite as unreasonable as he believes it to be. See for instance the (Grattan-Guinness 2008, p. 1), one of his points being that much of mathematics is developed specifically with the problems of the natural sciences in mind. This adds to the explanation of the level of intertwinement between the two.

The relation between the purely mathematical and (its applicability to) the physical world is often understood in terms of the existence of structural similarities between the two. One well-known account is the so-called 'mapping account' (Pincock 2012, pp. 27–29), which basic idea is that the mathematical model maps to the physical structure in a sense similar to a street map representing the layout of a city. Just as the map represents the city streets, the mathematical model represents the empirical phenomenon. But, as Bueno and Colyvan (2011, p. 348) rightly observe, such relation does not need to be isomorphic. Oftentimes there is more structure in the one than the other. For instance, there is less structure in the street map then in the city itself, as many aspects of the city are abstracted away from. But of course, any function relating the mathematical model to the physical phenomenon must retain the relevant structural relations *and* not add any unwanted structure to it.

One question that also needs to be clarified here, is what it is to say that the world has structure. The fact that a mapping appears to be possible does not in itself provide the answer. The remark by Maddy (1992, p. 282) we encountered in Sect. 2.1, about whether spacetime should be regarded as continuous because of the use of continuum mathematics, already anticipates this issue. Although the context in which she raised the question differs from the current one, the problem alluded to is related. Similiarly, Sklar (1974, p. 2) asks the question whether the geometric structure of the world, i.e. the geometric structure of spacetime, is a matter of convention or something that can be determined empirically?

If we look at what, generally speaking, structures are taken to be, then the world does not appear to possess those. At least not at first sight. Take for instance the characterization of what a structure is by Resnik (1997, pp. 202–203), who mostly prefers to speak of patterns rather than structures, and who takes them to be sets of objects on which relations between those hold. Bueno and Colyvan (2011) stress the fact that some such structure is either constructed or identified by our scientific theories. This is reminiscent of the distinction discussed by Shapiro (1997, pp. 84–90) between *ante rem* and *in re* structuralism. *Ante rem* structuralism, as the name suggests, states that the world has a structure that is independent of, or prior to, the theories we use to describe it. In line with platonism, the structure exists regardless of its instantiation. According to *in re* structuralism, on the other hand, the structure is not independent of, or prior to, the description of it. Because the structure is not considered to be independent of the its instantiation, getting rid of all instantiations also gets rid of the universal: "[g]et rid of all red things, and redness goes with them" (p. 84). Structuralist considerations will return in Part II, when the ramifications of a number theoretic counterpossible is discussed.

Regardless of any methaphysical questions about whether the world is carved up in a certain way or not, the point (Bueno & Colyvan 2011, p. 347) want to make is that for the mapping of a mathematical model onto the physical world to be possible, some structure of the wrold needs to be assumed, such as the streets in the toy example of the street map, in order to be able to map our mathematical models onto the world in the first place. This pre-theoretic, or 'assumed structure' as they call it, is used to kick-start the mapping. From here on, it can be refined according to empirical findings given a suitable interpretation, for instance by drawing inferences from the mathematical structure about the initial empirical setup (Bueno 2023, p. 50). So even though the mapping account might not be wrong, the challenge is to provide an account that can deal with the, often implicit, assumption that the world has a certain structure to which the mathematical model can be mapped.

This topic is a of course a very large one, and what is presented here is just meant to state some of the general ideas behind the objection that mathematics is explanatory, and that rather its role is, amongst others, representational.

3.3.2 Mathematics Merely Increases Expressive Capacity

The question is then what the contribution of mathematics is, if the strength of its contribution to extra-mathematical explanations, if explanatory at all, does not result in an increase in ontologically committing information? If it is indeed the case that the aspects of representation and explanation are not always separated correctly, perhaps because the physical interpretations of the mathematical formalisms are not always separated properly from the mathematical formalisms themselves, one suggestion is that mathematics helps us increase our expressive capacity in explanations (Knowles & Saatsi 2019).

Take for example the aspect of explanatory generality, as discussed in Sect. 3.1. If the relevant level of generality can also be reached by a nominalistically acceptable explanation (Sect. 4.2 will present such a strategy), then what stands out is the efficiency with which mathematics allows us to do so. Nominalist alternatives are oftentimes quite cumbersome. The way in which relevant information is presented using mathematics, and the ease with which it can be manipulated, makes it stand out. As Knowles and Saatsi (2019, pp. 1131–1132) put it with regard to the ease with which one can work with mathematics: "just plug in the appropriate numeral and calculate".

This increase of expressive capacity allows us to grasp an explanation, relative to a non-mathematical explanation, whithout compromising what Ylikoski and Kuorikoski (2010) call 'cognitive salience'. Cognitive salience is the idea that the ease with which, cognitively speaking, reasoning can be followed, implications can be recognized, and possible problems can be identified, stands out (Ylikoski & Kuorikoski 2010, p. 214).

The way in which an explanation is presented matters. If at some point we loose track of things due to a very high complexity in one explanation, but not in another, then this does indeed make the former a better explanation than the latter *in that respect*. This is not to say that it is necessarily the case that the more simplified an explanation is, the easier it can be grasped. Cognitive salience is not identical to simplicity, even though simplifying might often lead to an increase in cognitive salience. It is even so that, from a formal perspective, two explanations (or theories) can be equivalent, and thereby yield the same consequences, but one such explanation (or theory) can lead to a greater understanding of the subject matter (Ylikoski & Kuorikoski 2010, p. 215).

Ylikoski and Kuorikoski (2010, p. 205) make use of Woodward's (2003) interventionist theory of explanation, as discussed in Sect. 1.1, to determine the strength of an explanation with regard to understanding:

> [t]he degree of understanding conveyed by an explanation can now be defined as the number and importance of counterfactual inferences that the explanatory information makes possible.

The reference to counterfactual inferences here is particularly interesting in light of Part II, in which the role of counterfactuals in explanations will be further discussed, specifically in the context of mathematical explanations. What is important for now is that, given the inherent limited capacities we as cognizing beings have, the strenghth of an explanation is to be regarded relative to those limited cognitive abilities (Knowles & Saatsi 2019, pp. 1123). Mathematics helps us to utilize the limited resources by increasing our expressive capacity, in turn maximizing our understanding.

3.4 Inference to the Best Explanation

As was mentioned in Sect. 2.2, the EIA makes essential use of the abductive principle of inference to the best explanation, and therefore a brief discussion of a couple of the main objections to it is justified. The principle of IBE is widely accepted by scientific realists, who consider scientific theories to aim at truth. Given that some theory T is the best explanation of the available evidence, this is then considered to be a good reason to accept that theory. Subsequently, this should then also be a good reason to accept the existence of the entities postulated by that theory. As Lipton (2004, p. 1) nicely puts it: "our explanatory considerations guide our inferences".

Although most people involved in the discussions concerning the EIA accept the principle, be they from a nominalist or realist persuasion, in general not everyone thinks it is a sound principle. It is however a fundamental principle for the EIA, and it would not get off the ground if IBE were not accepted. A similar observation was made by Baker (2005, p. 255). Therefore, without the intention of nipping the argument in the bud here, it is worth mentioning some of the difficulties with it.

One of the main objectors of IBE is Van Fraassen, who staged a variety of arguments against it. One of the observations he makes, is that the theoretical virtues for selecting between different hypotheses might well be pragmatic rather than epistemic (Van Fraassen 1989, p. 201). This in itself already presents a difficulty for those defending the EIA, because the step from the comparison of different hypotheses to the conclusion that what one of those states *is* the case will then lack the right kind of justification. After all, pragmatics seems hardly in a position to adjudicate this kind of claims.

A different argument against IBE developed by Van Fraassen is known as the 'bad lot argument' (Psillos 1996, p. 36). The main aim of the argument is to show that the principle cannot be used as a rule for comparing different hypotheses in light of their explanatory success (Okasha 2000, p. 694). The observation by Van Fraassen is that the choice between theories rather is a choice between theories that have been formulated to date. In the selection between those theories, future possible theories are not considered (they cannot be considered, because they have not been thought up yet). The implication of this is that it is possible that none of the theories formulated thus far is (approximately) true. As a result, there is no ground for believing that the true theory is part of the set of theories under consideration. It could be the best explanation of a bad lot (Van Fraassen 1989, pp. 142–143). In fact, a prior belief about the likelihood of the true theory being in the current set is required:

> to take it that the best of set *X* will be more likely to be true than not, requires a prior belief that the truth is already more likely to be found in *X*, than not (Van Fraassen 1989, p. 143).

Thus, an additional premise is required in the argument in order to allow for the conclusion an IBE is supposed to provide.

The 'bad lot argument' spurred a lot of debate. Psillos (1996), for instance, criticizes the argument in relation as to how to interpret the remark by Van Fraassen that "our selection *may well* be the best of a bad lot" (Van Fraassen 1989, p. 143). He endorses an interpretation in terms of the likeliness of the truth of some hypothesis being outside the presently formulated set of hypotheses, i.e. that it is more plausible that the true hypothesis lies outside the presently formulated set of hypotheses (Psillos 1996, p. 37).

However, such an interpretation raises the question whether Van Fraassen could have meant something like that in the first place. For, if possibility of error were to be phrased in this way, then this is indeed too great a demand on the proponent of IBE: if there are many more ways in which the true hypothesis can be outside the formulated set of hypotheses to date, then this would mean that a proponent of IBE would have to argue that it is somehow nonetheless unlikely for the (approximately) true hypothesis to be outside the presently formulated set of hypotheses (Ladyman et al. 1997, p. 306). This is, as Psillos (1996, p. 37) rightly notices, a recipe for skepticism. After all, one observation is that in scientific reasoning, IBE appears to be a legitimate inferential practice. Ladyman et al. (1997, p. 306), the *et al.* including Van Fraassen, point out that what Van Fraassen rather asks is what the reasons a proponent of IBE could have for believing that the (approximately) true hypothesis is inside the present set of hypotheses, if not by assuming some form of privilege, i.e. an additional assumption that clarifies why it is to be believed that the right hypothesis is in the current set.

One possible response to the argument of the bad lot without having to assume privilege is suggested by Douven (2017). The suggestion is that a set can be created to capture all the possible explanations in logical space. Since this set is exhaustive, the best explanation surely is included. However, this newly formed set is hardly informative, for instance in terms of its empirical consequences. To derive any conclusions from it is hardly possible.

This leads to a second point of criticism, called the 'argument from indifference' (Psillos 1996, p. 43). Underlying the argument is some form of the principle of indifference, which roughly states that we know nothing about the truth value of the theories under consideration. As such, the argument of indifference states that the truth of some existing theory *T* is just as probable as the truth of any other competing theory (be they yet to be formulated or not, and perhaps infinitely many of them), but that the theories can disagree amongst them about many statements they contain *beyond* the evidence (Van Fraassen 1989, p. 146). This argument harbours the implicit assumption, argues Psillos (1996, p. 43), that all the knowledge we ultimately have about the theory under consideration, i.e. the theory of which some say that it explains the evidence or phenomenon best, is that it belongs to a class of a probably infinite number of theories which are equally explanatory. Again, this appears a recipe for skepticism, if no additional assumptions are allowed. Ladyman et al. (1997, p. 313) point out that even though IBE is frequently used in scientific reasoning, this in itself does not justify taking the principle for granted. If different explanations, which all agree in terms of the observations are possible, then referring to explanatory power is circular, and ontological questions cannot be decided on this basis. After all, the truth of the explanation needs to be assumed. This particular point will also play a role in Chap. 4.

To avoid such problems, another suggestion could be that the abductive step be phrased in terms of one hypothesis being more likely to be correct then the others (Douven 2017). This circumvents the problem of the inference of the absolute truth of some hypothesis. But then the question is what the comparison between the different hypotheses consists of. Clearly, a theory about closeness to truth is needed in this case.

3.5 Conclusion

This chapter discussed four often raised objections raised in the literature concerning the EIA. Section 3.1.1 stated the objection that mathematical explanations are better than nominalistic ones because of they are more general, and generality is often seen as one of the virtues of explanations. However, the objection stated that the scope of the generality used by proponents of the EIA goes beyond what is acceptable in a discussion within a naturalistic context. This point will come back in Chap. 4. The objection was stated by means of a counterfactual theory of explanation to bring out the explanatory relationship. This will return, although in a different guise, in Chap. 7.

Section 3.2 stated the objection pertaining to the possibility of multiple explanations. The objection concludes that, contrary to the original indispensability argument, it is not enough for the enhanced version to show the indispensability of mathematics, because of the abandonment of confirmation holism. It is claimed that IBE cannot adjudicate between the different available explanations, because it does not prevent allowing for inferring all of the explanations. Chapter 4 presents and alternative explanation, and also raises the problem of deciding which is the best explanation, in casu a mathematical and a nominalistic one.

Section 3.3 then presented an objection which states, against the EIA, that mathematics is not explanatory, but rather representational or indexing of the facts. The explanatory work is done by the facts. What is needed, is an suitable interpretation in order to relate the mathematics to the physical. Often, the distinction between explanation and representation is not made, thereby giving the impression that the mathematics explains. Looking at what the mathematical contribution in explanations then could be, it was suggested that it increases cognitive salience, thereby augmenting the possible scope of our understanding.

Lastly, Sect. 3.4 discussed some relevant objections to the principle of IBE in light of ontological conclusions, such as used in the EIA. Although IBE appears to be a valid inferential practice in the sciences, the question was raised whether this suffices for taking the principle for granted in arguments concerning the ontological status of mathematical objects.

Chapter 4
Circularity and Interpretability

Abstract The enhanced indispensability argument is at risk of being circular. In this chapter it is shown that paraphrasing the explanandum into first-order logic with identity only works when a bridge principle, i.e. Hume's Principle, is used to bridge the difference in context between the mathematical explanans and the translated explanandum. However, Hume's Principle must not be analytic, something which is not a given. An alternative 'interpretability strategy' is provided, which makes use of Robinson Arithmetic. Keeping in mind the broadly empiricist context of the argument, the difference in mathematical strength is discussed.

Having addressed some of the more common objections to the EIA in the previous chapter, this chapter focusses on another problem for the EIA, namely the problem of it being circular in a particular way. Section 4.1 presents the problem. First it introduces the problem as it was presented in the literature. Then the initial response to it is discussed. In this response, a possible solution is provided to escape the circularity by paraphrasing the explanandum into first-order logic with identity. By doning so, the intention is to avoid any reference to mathematical entities in the explanandum.

This is the point where the discussion is picked up. It will be argued that the proposed strategy cannot work, unless a bridge principle, i.e. Hume's Principle, is used to bridge the difference in context between the mathematical explanans and the translated explanandum. The use of Hume's Principle as a bridge principle has consequences as well. These consequences primarily pertain to the status of the bridge principle itself. Specifically, the question whether it has to considered to be an analytical principle or not becomes important. It will be argued that, if it indeed turns out that Hume's Principle is an analytical principle, then inevitably the explanandum will have to be interpreted as having mathematical content.

Taking into account that the problems the EIA faces as a result of the need for a bridge principle might not be considered a sound basis for an argument, Sect. 4.2 presents an alternative and nominalistically acceptable explanation making use of a so-called 'interpretability strategy'. In Sect. 4.3 it is shown that the alternative

© The Author(s), under exclusive license to Springer Nature Switzerland AG 2025

L. A. Tump, *On Mathematical Explanations of Empirical Phenomena*,
Synthese Library 517, https://doi.org/10.1007/978-3-031-98860-8_4

explanation has sufficient strength to explain the prime life-cycles of the Periodical Cicadas.

Regarding this point it is important to keep in mind that the EIA makes essential use of an inference to the best explanation (cf. Sects. 2.2 and 3.4). Here, there is also the risk that the argmuent becomes circular, although in quite a distinct way from the previously discussed circularity. It is accepted that the alternative explanation is not as strong as the mathematical explanation. However, if in choosing the best explanation the proponents of the EIA argue that because the mathematical explanation is stronger it is therefore better, it is noted that this begs the question in the broadly empiricist context in which the discussion takes place.

The chapter ends with some concluding remarks about the, perhaps somewhat unexpected, relevance of certain discussions external to the present debate in Sect. 4.4.

4.1 The Problem of Circularity

Analyzing the example of the mathematical explanation provided in Sect. 2.4, Bangu (2008) observes that the explanandum, at least in the way it is presented in its original form,

> consists in a property-attribution claim, where the property in question is primeness, a paradigmatic example of a mathematical property (Bangu 2008, p. 18).

The mathematical property in the original 'why'-question related to the life cycles of the Periodical Cicadas is indeed primality, which is a property typically predicated of numbers. As mentioned earlier, the problem with explananda containing mathematical *entities* is that, and we remind ourselves that the strategy of inference to the best explanation is accepted by at least some of the proponents of both sides in the debate, the explanandum must be assumed to be true. As Leng (2005, p. 174) puts it:

> [g]iven the form of [the] argument, one might wonder why it is mathematical explanations of physical phenomena that get priority. For if there are ... some genuine mathematical explanations ... then these explanations must also have a true explanans. The reason why this argument can't be used is that, in the context of an argument for realism about mathematics, it is question-begging. For we also assume here that genuine explanations must have a true *explanandum*, and when the explanandum is mathematical, its truth will also be in question.

Thus, for there to be something to explain at all, the truth of that which is to be explained must be assumed. The existence of some empirical phenomenon, of which it is thought it needs to be explained mathematically, was after all the reason why such an effort was made to find an example of an extra-mathematical explanation. Although the explanandum in the EIA, i.e. why the life cycles of the Periodical Cicada are prime, does at first sight not appear to contain any mathematical entities, Bangu (2008, p. 18) notices that

> if mathematical properties apply to anything, they apply prima facie to mathematical objects.

In other words, given the presence of the mathematical property of primality in the explanandum, it therefore *does also* contain a reference to a mathematical entity. Primality, as mentioned before, is typically predicated of numbers. As a result, given the assumption that the explanandum must be true because of the use of the IBE, it would be the case that one is ontologically committed to mathematical entities after al, namely numbers. But then it would be the case that mathematical platonism is established before arguing for it, and as such the argument would indeed be circular.

4.1.1 Response to the Claim of Circularity

As Baker (2009, p. 620) rightly notices, the claim of circularity has some force. He acknowledges the fact that the way the example was initially formulated was indeed problematic due to the indirect reference to mathematical entities, i.e. numbers. To escape the allegation of circularity, he therefore reformulates the argument so as to prevent the explanandum containing a mathematical property and hence the explanandum from having the object(s) to which the property applies. The basic structure of the argument is now presented with the explanandum no longer mentioning the property of primality. Instead of the primality of the life cycles of the Periodical Cicada, what is to be explained now is the actual length of the life cycles in years (Baker 2009, pp. 620–622):

(D) **Data:**
The length (in years) of the life cycle of cicada subspecies A is 13.
The length (in years) of the life cycle of cicada subspecies B is 17.

As can be seen, the data now mentions mathematical entities directly, in casu numbers. This might seem odd, because the reason for the reformulation was to get rid of the reference to mathematical entities. But one thing that can be done with the numbers mentioned in the data (D), it is suggested, is to paraphrase them away in first-order logic with identity. This way, one gets rid of the reference to the mathematics in the explanandum entirely. First-order logic with identity is after all not generally considered to be mathematical in nature.

With this in mind, the first datum, with F as the predicate for 'year of the life cycle of cicada subspecies A', paraphrased in first order logic with identity (*fol*) is:[1]

[1] For completeness, both the 13- as well as the 17-year life cycles were mentioned in (D). However, to prevent too much repetition, in what follows only the first datum will be used. A similar paraphrase can of course be made for the 17-year life cycle.

(D_{fol}) **Data *fol*:**

$$\exists x_1 \ldots \exists x_{13}(Fx_1 \wedge \ldots \wedge Fx_{13} \wedge x_1 \neq x_2 \wedge \ldots \wedge x_{12} \neq x_{13} \wedge \forall x_{14}(Fx_{14} \leftrightarrow (x_{14} = x_1 \vee \ldots \vee x_{14} = x_{13})))$$

It can be clearly seen from the above that the reason for the shift from an explanandum containing the notion of prime to an explanandum containing numbers, is that numbers can be formulated in first-order logic with identity, whereas the notion of prime cannot be paraphrased away in first-order logic with identity. Baker (2009, p. 619) of course recognized this. The reason why the notion of 'prime' in a sentence like "The length (in years) of the life cycle of cicada subspecies A is prime" cannot be expressed in first-order logic with identity, is because of the infinite number of ways in which a number can be prime.[2] Any expression in first-order logic with identity intended to paraphrase such a sentence would therefore need to make use of an infinite disjunction.

The notion of primality is however still needed for the explanation of the prime life cycles of the Periodical Cicadas. To avoid circularity, the reference to it is therefore put tentatively as an hypothesis in (H). Putting it tentatively is necessary, because the hypothesis *does* refer to this mathematical property, and therefore implicitely contains mathematical entities. The tentative hypothesis will ultimately be used to 'bootstrap' the explanation of data (D) in relation to the observed primality of the life cycles. Given these changes to the argument, it continues as follows:

(H) **Tentative hypotheses:**
The length (in years) of the life cycle of cicada subspecies A is prime.
The length (in years) of the life cycle of cicada subspecies B is prime.

(1) **Biological 'law'**
Having a life cycle period which minimizes intersection with other periods is evolutionary advantageous.

(2) **Number-theoretic theorem**
Prime periods minimize intersection (compared to non-prime periods).

(3) **Explanation (E)**
The lengths (in years) of the life cycles of periodical organisms are likely to be prime (explanation of (H)).

(4) **Ecological constraints**
The subspecies of Periodical Cicadas are limited by their respective biological constraints.

(5) **Extended explanation (E*)**
(E) with added biological constraints yields an explanation of (D).

---

[2] This is due to the Löwenheim-Skolem theorem (for the downward and upward theorems respectively, see (Boolos et al. 2007, p. 147,163)). Because of the infinite model, the upward theorem states that there are nonenumerable models. These are non-standard models, and as such it is not possible to state that an object is a natural number in first-order logic. See also Boolos (1981, p. 466).

Steps (1) through (5) correspond to the initial argument structure given in Sect. 2.4, with steps (H) and (1) through (3) being the tentative part of the argument. The remainder of the argument is an inference to the best explanation, which is used to establish the ontological commitment to numbers and their properties, in this case the property of them being prime.

(6) (E*) is the best explanation of (D).
(7) Hence we ought to believe (E*), and thereby (E).
(8) But (D) and (E) together imply (H).
(9) Hence we ought to believe (H)

The basic idea underlying the reformulated argument structure is that, by paraphrasing data (D) in first-order logic with identity, and putting the hypotheses pertaining to the primality of the periodical cicadas life cycle tentatively, circularity can be prevented.

Regarding the tentative hypotheses in (H), there are two possibilities. The primality of the length (in years) of the life cycles could either have a common explanation, describing them as prime, or have no common explanation, in which describing them as prime is a coincidence (Baker 2009, pp. 621–622).

It is subsequently argued that the common explanation is the better one, because of theoretical virtues like generality and modal strength,[3] and that hence we ought to be committed to the life cycles being prime. So, the suggestion is that the *best* explanation of tentative hypotheses (H) is explanation (E). And, given that very fact, it is then claimed that it

> can *also* be turned into an explanation of ... [(D)], by adding the subspecies-specific ecological constraints (Baker 2009, p. 621).

This should warrant the conclusion of the EIA, i.e. that the life cycles are prime, and that we are ontologically committed to the existence of numbers.

However, this is not all there is to say about the internal workings of the reformulated argument. With the paraphrasing of the reference to numbers in the explanandum, the context of the sentence has changed. That is, the context has shifted from numbers in the original version to first-order logic with identity in the new verion. This raises the question of the connection between the explanandum and the explanans in the latter vesion. How can an explanation in which the mathematical property of prime is used, such as in the mathematical explanation (E) of tentative hypothesis (H), be turned into, or transferred to, an explanation of a datum which is paraphrased away in first-order logic with identity, such as in (D_{fol})? This will be the topic of the next section.

[3] The use of these theoretical values by mathematical platonists in order to argue for certain explanations to be the best explanations is not uncontroversial, as we have seen in Chap. 3. The aspect of the greater strength of mathematical explanations over their non-mathematical counterparts will also be relevant in Sect. 4.2.

4.1.2 The Circularity Problem Again

The main argument of this section questions the feasibility of (E*) yielding an explanation of (D) within the context of the EIA. To be specific, the inference from '__ is (likely to be) prime', which is used in (E*) because of (E), to '__ is 13', which is used in (D), needs explaining.

Assuming that the mathematical explanation of the tentative hypotheses, which states that the life cycles are prime, is accepted as the best explanation, the next step is to explain how the explanation of the tentative hypotheses can be turned into an explanation of data (D). Recall that (D) was paraphrased in first-order logic with identity, as we have seen in (D_{fol}). Note that, on the one hand, datum (D_{fol}), treats of a *plurality*, that is, the collection of objects expressed by the sentence in first-order logic with identity. But explanation (E), on the other hand, treats of an *entity*, which (likely) has the property of being prime. As a result, the context of the two sentences differs. The context in the explanandum after its reformulation has shifted from entitties to pluralities. As a result, because primality relates to entities, it is not possible for the explanans to predicate primality of the datum in first-order logic with identity. Therefore, for the newly formulated explanation to be possible, the two contexts need to be linked up somehow.

The crucial observation here is that there is a gap between (D_{fol}) and (E*): how can the plurality of (D_{fol}) end up as an entity about which (E*) claims that it is (likely) prime?

Taking this observation into account, and thus for Baker's move to prevent circularity to work, we need some sort of bridge principle, in order to be able to logically relate the explanans to the explanandum by connecting entities to pluralities. Without such a logical connection, it is not possible for there to be a mathematical explanation of the empirical phenomenon described by the sentence in first-order logic with identity.

In light of Frege's work (Frege 1893) and the neo-Fregean research program (Burgess 2005), a natural suggestion would be to use Hume's Principle (HP) as a bridge principle here, which states that "for any concepts F, G, the number of F is identical to the number of G if and only if F and G are equinumerous" (Shapiro 2000, p. 110):[4]

HP $\forall F \forall G(\#F = \#G \leftrightarrow F \cong G)$

As can be seen, this principle connects the pluralities, i.e collections of objects, having properties F and G to entities (which in this case will be numbers).

In order to determine in which cases the concepts F and G are equinumerous, a relation R needs to be defined. The relation R must be a bijection, mapping each element of the one set exactly to one element of the other set. If this relation R obtains between sets, then these sets are said to be equinumerous. The definition of equinumerousity can now be stated as follows (Burgess 2005, p. 24):

[4] The notation is from (Boolos 1998, p. 139).

$$F \cong G \leftrightarrow \exists R(\forall x(Fx \rightarrow \exists! y(Gy \wedge Rxy)) \wedge \forall y(Gy \rightarrow \exists! x(Fx \wedge Rxy)))$$

Equinumerousity is an equivalence relation: it is reflexive, symmetric, and transitive. Equivalences between sets can be 'abstracted'. Abstraction in this sense means that an entity can be assigned to (a collection of) objects. This abstracted entity will then be identical for two (collections of) objects if and only if these (collections of) objects are equivalent. For the abstraction the so-called 'cardinality operator' (#) is used, which can be read as 'the number of', and is associated with a concept F or G. The operator thus assigns the cardinal number which designates the cardinality of the set of the objects falling under some concept of this set. This means that, when the concept concerned is F, $\#F$ reads 'the number of F', and assigns the cardinal number which designates the cardinality of the set of the objects falling under this concept F (cf. (Burgess 2005, pp. 22–24)). For example, if F would stand for the concept *planets of the solar system*, then a statement like '$\#(planets\ of\ the\ solar\ system) = 8$' can be formulated, in which the cardinal number 8 is assigned to the set of planets of the solar system.

If HP is subsequently combined with dyadic second-order logic,[5] then Frege's Theorem says that the Peano axioms can be deduced, given Frege's definitions of zero, successor and natural number (Burgess 2005, p. 26). This is important, because the Peano axioms define the natural numbers. Next, in order to be able to say something about prime numbers, arithmetical operations such as addition and multiplication are needed. According to Dedekind's Theorem, the existence of unique operations on natural numbers satisfying the recursion equations, or definitions, for addition and multiplication can be deduced in dyadic second-order logic from the Peano axioms (Burgess 2005, p. 29). With the natural numbers and the arithmetical operations on them available, it is possible to start proving results about prime numbers.

This looks very promising as a start, but note however that, given HP and Frege's above mentioned definitions, (D_{fol}) is provably equivalent with the following statement:

$D_{\#}$ $\quad \#F = 13$

Now, the question is whether this provable equivalence is also an analytical equivalence. In the neo-Fregean context, what is traditionally meant by 'analytical equivalence' is that the equivalence follows from nothing but logic and definitions (Wright 1999, p. 9). The question whether $D_{\#}$ is an analytical equivalence is in fact important for determining if the reformulated EIA is circular or not. This is so, because if HP turns out to be an analytical principle, then (D_{fol}) is thereby analytically equivalent with $D_{\#}$. And therefore, given Frege's Theorem and Dedekind's Theorem, (D_{fol}) has to be interpreted as having mathematical content, because content from the one side is carried over to the other side. This would render the EIA circular.

---

[5] Second-order logic is needed, because besides quantifying over elements of sets (e.g. numbers), we also need to be able to quantify over relations or predicates (e.g. primality). The dyadic variety, as opposed to monadic, allows for the use of binary predicates.

Since the only things that have been assumed so far are dyadic second-order logic, Frege's definitions, and HP, the question at this point really only boils down to the question whether HP itself is an analytical principle.

It is clear that HP is at least a material equivalence statement. But neo-Fregeans will immediately point out that HP is not just a regular material equivalence. Take for instance the following example, borrowing from Quine (1961, p. 37). The material equivalence

'creature with a heart' if and only if 'creature with kidneys'

does not carry over any content from the one side to the other. This is because the material equivalence does not hold out of *necessity*. Rather, it holds by virtue of accidental matters of fact. Creatures with hearts happen to also be creatures with kidneys. But HP on the other hand functions, at the very least, as an axiom for $\#F$, and is therefore, trivially, a provable equivalence. But what is more, it is taken to be an implicit, or contextual, definition, thereby allowing for the cardinal numbers to be defined. As Wright (1999, p. 10) puts it:

> [t]he key idea [of HP in the neo-Fregean context] is that an instance on the left-hand side of an abstraction principle is meant to embody a *reconceptualization* of the type of state of affairs depicted on the right.

And precisely because of this reconceptualization, HP is often taken to be an analytic principle. If the principle is considered in this light, then (D_{fol}) is indeed analytically equivalent to $D_{\#}$. Consequently, in the explanation (i.e. the ordered set of the explanandum and the explanans), the explanandum is also analytically equivalent to a sentence that has to be interpreted as having mathematical content. And as a result thereof, the explanandum then also has to be interpreted as having mathematical content. But then the argument would beg the question against the nominalist, just as it did in its original form, i.e. before the reformulation of data (D). Thus, if it turns out that HP is indeed an analytic principle, the proposed strategy of paraphrasing will not prevent the argument from being circular.

The question concerning the analyticity of HP is not settled. And more so, not all negative answers to it result in a vindication of the use of HP in the EIA. Boolos (1998) for example, has argued that HP should not be considered to be an analytical principle, but for reasons which would complicate matters even more for a position such as Baker's. The problem is that HP, if it were to hold, entails the existence of infinitely many abstract objects. And analytic principles should not commit to so many objects, let alone abstract objects. Rather, analytic principles should be ontologically neutral. That is, they should not make any substantive claims (Boolos 1998, pp. 303–306).

In response to the criticism formulated by Boolos, some have argued that questions of meaning and questions of existence should be separated. The existence of numbers does not follow from HP, or so it could be argued. Rather, the existence of numbers is presupposed, resulting from the fact that neo-Fregeans consider HP to be true.

If questions of existence were indeed, *pace* the neo-Fregeans, to be settled elsewhere, this would mean that HP would therefore not commit to the existence of numbers, and could perhaps still be considered an analytic principle. However, such a move would not work for proponents of the EIA, because the argument itself would then be neutral with regard to the existence of numbers. In fact, this is why a conditional version of HP is acceptable for a nominalist like Field (1984, p. 661):

HP$_{\text{cond}}$ If numbers exist, then $\forall F \forall G(\#F = \#G \leftrightarrow F \cong G)$

Clearly, HP$_{\text{cond}}$ would only bridge the gap between arithmetic on the one hand and (D_{fol}) on the other hand conditional upon the existence of numbers. So, we would only have a mathematical explanation if it is already agreed upon that numbers exist. But the EIA was supposed to justify the belief in the existence of numbers in the first place. Circularity looms again.

Wright (1999, pp. 9–10) points out that, in effect, the intelligibility of HP$_{\text{cond}}$ itself presupposes that numbers exists. The antecedent 'numbers exist' will contain expressions of the form $\#X$. But there is no longer a sufficient condition to determine the truth of sentences containing those expressions, and therefore, there is no longer a contextual definition available for those. How is the antecedent to be understood? This is why he concludes that HP$_{\text{cond}}$ has to be "relativized to an *antecedently given* domain of quantification" (p. 9) and that "the existence of numbers is a *further* fact, something which the (mere) equinumerousity of concepts may leave unresolved" (p. 10). If analyticity is supposed to be ontologically neutral, then the analyticity of HP$_{\text{cond}}$ cannot be accepted.

This, then, is the problem of circularity the EIA faces if it is intended as an argument for mathematical platonism: for there to be a mathematical explanation of the first-order logic datum at all, the mathematical explanans needs to be able to connect to the explanandum. However, resulting from the strategy to paraphrase the reference to numbers away in first-order logic with identity, the explanandum now contains a plurality, resulting in a difference between the two cases.

To overcome the difference in context, HP is a natural suggestion to function as a bridge principle, because it can be used in combination with dyadic second-order logic to say things about prime numbers. This is relevant, because without HP (or some other bridge principle connecting entities to pluralities) there would be no logical connection between the plurality in the explanandum and the property of primality in the explanans, and thereby no mathematical explanation. But the predicate '__ is prime' can either not latch on to the first-order logic sentence without the latter being interpreted as a mathematical statement, resulting from the supposed analyticity of HP, or the predicate can not latch on to the first-order logic sentence with identity without making it conditional upon the very existence of numbers, as would be the case with a conditional version of HP.

Of course, much depends on the status of HP, i.e. whether it is indeed an analytic principle or not. There is an ongoing debate about this. Recently, for instance, Boccuni and Woods (2018) defend the idea that HP is a conceptual truth, whereas Darnell and Thomas-Bolduc (2018) argue that HP is not analytic. The outcome of that debate is relevant for the assessment of the EIA. What *is* clear, is that, in the end, HP must be a non-analytic principle for the EIA to work.

One last remark with regard to requirement that HP be a non-analytic principle. Baker (2021, p. 158) suggested, in a response to the article which is the source of this chapter, for HP to be included in the bootstrap procedure, i.e. adding it to the overall explanatory inference. The idea is that, just as the reference to primality was at first only included tentatively, so will HP be included tentatively at first.

In the case of the primality of the number of years of the life cycles, it was argued that this was the best explanation of the tentative hypotheses in (H). And therefore, the explanation of the primality of the number of years of the life cycles could also be turned into an explanation of the nominalistically acceptable data in D_{fol}. In the case of HP, the suggested idea is that if the principle is also put tentatively at first, and that ultimatley it turns out that it is also part of the best explanation of D_{fol}, for example because it is needed to bridge the gap in context between the paraphrased sentences in D_{fol} and data (D), then this does not render the argument circular either and the EIA can go through even in the case of HP being analytic.

However, as discussed earlier in this section, the issue here is that if HP is an analytic principle, then mathematical content is carried over to the explanandum. This would also be the case if it is inferred, tentatively at first, that HP together with the common explanation of the primality of the number of years of the life cycles (are part) of the best explanation of the Periodical Cicada's life cycles. To be sure, the paraphrasing of the number talk in data (D) was precisely meant to prevent mathematical content to be present in it. With the addition of HP to the argument in order to bridge the difference in context, D_{fol} still has mathematical content. Thus the fact remains that in the case that HP turns out to be an analytical principle, mathematical content carries over to the explanandum paraphrased in first-order logic with identity, because both sides of the material equivalence are not independent in that respect. This will remain the case if HP is added to the tentative part of the argument, because it would be inherent to the principle itself if it turned out to be analytical. The point is quite subtle indeed. But it is clear that in predicating primality of the plurality in datum D_{fol} cannot be done in a manner independent of the question of the analyticity of HP, whether it was first put tentatively or not.

The current state of where we currently are can be briefly stated as follows. In the previous section a gap in the EIA was identified and this gap was bridged with the help of HP. It was then argued that the circularity problem resurfaces, if HP is an analytic truth. Thus, for the bootstrapping strategy as suggested by Baker to work, and to allow for the possibility to escape circularity, the EIA has to assume that HP is non-analytic. The non-analyticity of HP is not something which is a matter of course, and still an ongoing discussion.

In the next section it will nonetheless be supposed, for the sake of argument, that HP is not an analytic principle. The defender of the EIA could then give the following explanation: HP with dyadic second-order logic and added biological constraints yields an explanation of (D_{fol}). If it turns out that this is the best explanation, then the conclusion of the indispensability argument is that one should accept the existence of (Frege-)numbers.

Although the explanation would not suffer from circularity this way, it has not yet been settled whether the mathematical explanation based on HP is the *best* explanation, which is needed for the EIA to establish platonism. The next section challenges the claim that the EIA is indeed the best explanation, thereby arguing that even if it would turn out that HP could be considered to be a non-analytic principle, hence allaying the worries of circularity, platonism would still not be established.

The plan for the next section is as follows: while assuming the non-analyticity of HP, an alternative explanation strategy, called 'the interpretability strategy', will be provided for the opponent of the EIA, which will make use of relative interpretability theorems. Section 4.3 will then show how this can be applied to the number-theoretic example. In line with the concern of Sect. 4.1, the circularity concern will be picked up once more, and it will be pointed out that one criterion for the selection of the best explanation, namely *mathematical* power, is question-begging for a nominalist.

4.2 The Interpretability Strategy

In the context of debates on the indispensability of mathematics to science, it is interesting to find out what the weakest theories are that can interpret the mathematics that is needed in scientific explanations. On the one hand this has intrinsic value, but on the other hand, it is interesting to see what natural phenomena can be explained nominalistically, specifically those of which it is claimed that their best explanations are in fact mathematical. Also, finding out what the weakest theories are for interpreting the mathematics used in scientific explanations can shed light on what *kind* of mathematics appears to lend itself for such weak interpretations.

The mathematical explanation of the prime life cycle of the Periodical Cicada, stating that prime life cycles are advantageous because they intersect less with other life cycles than non-prime life cycles, rests on the number-theoretic lemmas that the lowest common multiple of two numbers is maximal if and only if they are coprime, and that two numbers are coprime if and only if they are prime (for details, see for instance part 1 in Landau 1958).

To see whether these notions can be dealt with without number theory, let us have a look at an alternative explanation strategy. This strategy will assume some aspects of intermediate logic—see, for instance, Boolos et al. (2007). Also, the notion of relative interpretation will also be made use of. Roughly, a relative interpretation of a theory into another theory is a translation of the language of the first theory into the language of the second theory. The translation has to meet certain conditions, such as the preservation of the definitions of constants, functions, or relation symbols from the first theory in the second. It also requires a proof that the translation of every theorem of the first theory is in fact a theorem of the second theory. For formal definitions, see (Enderton 2001, section 2.7).

The starting point of the interpretability strategy is a theory of concatenation of non-empty strings of a's and b's, called QT, which is a first-order language, and consists of identity, individual constants a and b, a binary function symbol $*$ for the concatenation of strings, and the following five[6] axioms (Damnjanovic 2017, pp. 383–384):

QT1 $((x * y) * z) = (x * (y * z))$
QT2 $\neg(x * y = a) \wedge \neg(x * y = b)$
QT3 $(x * a = y * a \rightarrow x = y) \wedge (a * x = a * y \rightarrow x = y) \wedge$
$(x * b = y * b \rightarrow x = y) \wedge (b * x = b * y \rightarrow x = y)$
QT4 $\neg(a * x = b * x) \wedge \neg(x * a = x * b)$
QT5 $x = a \vee x = b \vee (\exists y(a * y = x \vee b * y = x) \wedge$
$\exists z(z * a = x \vee z * b = x))$

The result is a theory about the concatenation of non-empty strings of a's and b's.

The idea is that QT is an interesting starting point for a nominalist. First, Goodman and Quine (1947) already made use of expressions of concatenation in their nominalistic theory of syntax, which they use in an attempt to reconstruct classical mathematics. In fact, one could say that any metaphysical theory that claims to be true should be compatible with a theory of truth of sentences. Such a theory depends on a theory of syntax, which in turn is hardly conceivable without some kind of theory of concatenation. See for instance (Tarski 1936). Goodman and Quine (1947, p. 106), however, decline to assume that there are infinitely many objects. As it will turn out, QT has no finite models. So, QT goes beyond what they are willing to assume. However, other nominalists, notably Field, have no scruples in assuming that there are infinitely many objects.

Second, it turns out that QT is mutually interpretable with Adjunctive Set Theory (AST) with Extensionality (EXT)—see Damnjanovic (2017). The language of AST+EXT is a first-order language with binary relation symbol $\in$. The theory has only three axioms:

NULL $\exists x \forall y \neg y \in x$
ADJ $\forall x \forall y \exists z \forall w (w \in z \leftrightarrow (w = x \vee w = y))$
EXT $\forall x \forall y (\forall z (z \in x \leftrightarrow z \in y) \rightarrow x = y)$

The result is a very weak fragment of Zermelo-Fraenkel set theory with Choice (ZFC). Field (1992) has tried to show that ZFC with so-called *ur*elements, i.e. an object that can belong to a set but is itself not a set, is 'conservative' in the sense that it is consistent with any internally consistent purely physical theory and adds nothing (non-mathematical) to it.[7] This played an important role in

[6] An additional axiom can be added for the successor function, **QT6** $Sx = y \leftrightarrow ((x = a \wedge y = b) \vee (\neg x = a \wedge x * b = y))$, with the system of the combined 6 axioms called QT$^+$. The resulting system is however a conservative extension of QT, because the added axiom is essentially a definition (Damnjanovic 2017, p. 384).

[7] See also (Boyce 2020, pp. 13–14) or (Colyvan 2001, pp. 70–72) for brief discussions on the conservativeness of mathematics and the relation to it being suitable for the sciences.

his nominalization project of physics. And because even Field thinks that ZFC with *ur*elements is conservative, AST should then most certainly be considered as conservative.

The interesting thing is that QT is mutually interpretable with Robinson arithmetic, denoted by Q (Damnjanovic 2017). The language of Q is as follows: $\mathcal{L}_Q = (0, S, +, \cdot, =)$, with 0 a constant, S the function symbol for successor, $+$ and $\cdot$ the function symbols for addition and multiplication, and $=$ the symbol for the equality relation. It consists of the following axioms (Tarski et al. 1953, p. 51), with the first three defining the function S and the initial element 0:

Q1 $(Sx = Sy) \rightarrow (x = y)$
Q2 $0 \neq Sy$
Q3 $(x \neq 0) \rightarrow \exists y(x = Sy)$

The first axiom states that the identity of successor implies the identity of number, or put negatively, different numbers have different successors. The second axiom states that there is an element, 0, which is not the successor of any number, i.e. it is the initial element. The third axiom states that every number, except 0, is the successor of some other number.

The remaining four axioms define the recursive definitions of addition and multiplication (Tarski et al. 1953, p. 51):

Q4 $x + 0 = x$
Q5 $x + Sy = S(x + y)$
Q6 $x \cdot 0 = 0$
Q7 $x \cdot Sy = (x \cdot y) + x$

For convenience, Q is often extended to include the order symbol $<$, such that $\mathcal{L}_{Q^<} = (0, S, +, \cdot, =, <)$, which in turn requires the addition of the following axiom (Boolos et al. 2007, p. 216):

Q8 $x < y \leftrightarrow \exists z(Sz + x = y)$

The resulting system $Q^<$ is however conservative with respect to Q.

Although it is good news that QT is mutually interpretable with this very weak system of arithmetic, it might at first sight seem disappointing that it can *only* interpret such a weak system of arithmetic. This is especially so when compared to second-order Peano Arithmetic (PA^2), which one can get out of the combination of HP and dyadic second-order logic.[8] But it is important to keep the dialectical situation in mind. The interpretability strategy is not meant to give a nominalistic

[8] Interestingly, Q is not interpretable in mereology, which is also often favoured by nominalists. This is a consequence of the following theorem listed by Niebergall (2011): there is no consistent mereological theory in which AST is relatively interpretable. This is certainly relevant for the discussion about the evaluation of the nominalistic paraphrase of statements about primality given by Tallant (2013), because he uses mereological fusions. Taking it further, he hopes that mereology will be enough to paraphrase all "scientifically explanatory mathematics". The latter cannot include Robinson Arithmetic, if his project is to succeed.

reconstruction of the totality of mathematics, or even a reconstruction of as much mathematics as is possible. Neither is it preoccupied with developing a nominalistic reconstruction of all the mathematics that is used in natural science. The main point of interest here, is to see whether there is a nominalistic explanation for those natural phenomena for which it is claimed that they are best explained mathematically, in casu number-theoretically.

4.3 Interpreting Number Theory

Because our example explanation of the life cycle of the Periodical Cicada makes use of a number theoretic theorem, a very weak system of arithmetic like Q is interesting to look at, because it has sufficient strength to explain the prime life cycles of the Periodical Cicada. It does so, however, with requiring very little resources. Theory Q has some deductive power too: for any Σ_1^0-sentence ϕ, ϕ is provable in Q if and only if ϕ is true in the standard model of arithmetic. With a Σ_1^0 is meant an existential formula of the form $\exists x \psi$, with ψ a formula built up from atomic formulas using only the logical operators and bounded quantifiers. The latter are of the form $\forall x \leq t$ or $\exists x \leq t$ (with t a term not containing x). Moreover, every recursive function and relation is definable in Q by a Σ_1^0-formula.

The question is now whether Q has *enough* deductive power for providing a mathematical explanation of the cicada case? First, note that the multiplication and quotient functions, the maximum and minimum functions, and the properties of being prime and co-prime belong to the categories of recursive functions and relations. Therefore, we can formulate all the necessary key notions with the help of Σ_1^0-formulas.

Second, one may worry that the lemmas used by Baker are much stronger, for they are unbounded universal quantifications ranging over *all* numbers, and not just the numbers smaller than some number. However, those unbounded universal generalizations are far stronger than needed in the explanations of natural phenomena. Take the case of the cicadas. There is no reason to consider life cycles of cicadas that exceed, let's say, the life span of the Earth. There is no problem with using bounded universal quantifications, because whatever natural phenomenon that needs to be explained is going to be bounded anyway. Recall, for example, the ecological constraints that were used in the (reformulated) EIA. It is exceedingly likely that similar constraints will be used in any purported mathematical explanation in natural science. Remember Sect. 3.1 for the similar observation by Knowles and Saatsi (2019, p. 9) in the context of a counterfactual theory of explanation. The possibility of extreme values (in counterfactual situations) due to the limitless scope of the mathematics invoked in an explanation is far removed from the actual theoretical context for which scientific theories generally are developed. In order to be able to evaluate such extreme counterfactuals, what they call 'super-empirical issues of modal semantics' are required.

Baker (2016) also offers a 'more general' explanation of the primality of the life cycles of cicadas which is based on Bertrand's postulate. The interpretability strategy can be applied to this explanation too, as the relevant notions can be formulated in a similar way. In fact, the interpretability strategy sketched above can actually be applied in other kind of cases as well. The reason is that Q can interpret in its turn a quite wide range of mathematical theories, including (first-order) Euclidean geometry, consequence, Tarski's theory of real closed ordered fields, and the so-called Base Theory for Feasible Analysis (Ferreira & Ferreira 2013).

This means that in all cases where explanations essentially involve any of the above mentioned theories, one can use the relative interpretability theorems to conclude, first, that there are theorems in Q that are interpretations of the theorems in the aforementioned theories and, second, that there are theorems in QT that are interpretations of the relevant theorems in Q. This is the core of the so-called 'interpretability strategy'. Metaphorically speaking, what is proposed is to run simulations of software written in one language within other software written in another language, but which is ultimately reducible to the manipulation of bits. Ontologically speaking, the nominalist is committed to no more than non-empty strings of *a*'s and *b*'s, because that is what the existential quantifiers in QT range over.[9]

Independent of ontological controversies, the thought here is also that the interpretability strategy is interesting in the context of debates on mathematical applications and mathematical indispensability for science. After all, it is worthwhile to know what the weakest mathematical theories are in which one can interpret whatever mathematics is used in scientific explanations.

In this sense, the strategy is somewhat similar to a strategy suggested by Bueno (2001, p. 119). His idea was for a nominalist to capitalize on the so-called *reverse mathematics* program, initiated by Harvey Friedman, and aiming at discovering what the weakest axioms are to prove a theorem. There are a couple of differences to keep in mind, though. One is that Bueno is aiming at a nominalistic reconstruction of as much of mathematics as possible, whereas the aim here is much more modest, namely that we are only talking about what Tallant (2013, p. 2074) calls "scientifically explanatory mathematics". Another difference is that Bueno focuses on abstraction principles, which he wants to paraphrase using a combination of plural logic and mereology.

[9] We are relying here on Quine's criterion for ontological commitment (Quine 1948). As we have seen in secions 2.1 or 3.3, some participants in the debate have questioned this criterion—see also, for example, Azzouni (2004).

4.4 Conclusion

Section 4.2 stated that there are at least two alternative explanations available in the cicada case. The first is an explanation invoking HP (of which it also must be assumed to be non-analytic) and dyadic second-order logic, which together form a theory called 'Frege Arithmetic'. Additionally there are some ecological constraints. The second is an explanation invoking Robinson Arithmetic, which is subsequently interpreted within a theory of concatenation of non-empty strings of a's and b's. If one intends to conclude that (Frege-)numbers exist, which is in line with platonism and which goes against nominalism, it needs to be established that the first explanation is indeed the best explanation. This has to be done without begging the question. Section 4.3 observes that Frege Arithmetic has a lot more mathematical strength than Robinson Arithmetic. But a nominalist should not necessarily be concerned about that. A difference in empirical scope, however, would be cause for concern for a nominalist. This is a challenge for the defender of the EIA.

Of course, as discussed in Sect. 4.1, some philosophers think that HP is *not* an analytic truth. In that case there can be a mathematical explanation of the explanandum, paraphrased in first-order logic. This explanation makes use of HP, dyadic second-order logic and certain ecological constraints. But even if the EIA could go through this way, it is not yet settled that this is the *best* explanation.

The suggested interpretability strategy shows this. With the possibility of using the relative interpretability of several mathematical theories into an elementary theory of arithmetic, called 'Robinson Arithmetic', in turn interpretable in a theory of concatenation of non-empty strings of a's and b's, there is a possible alternative. This alternative is indeed somewhat weak, but strong enough for what is required to interpret the theorems of quite range of mathematical theories, and as such the purported mathematical explanations that make use of these theories. Sure, the mathematical power of Robinson Arithmetic is much less than the mathematical power of Frege Arithmetic, but for a nominalist that is a question-begging difference, whereas a difference in empirical scope would not be question-begging.

With regard to the difference in ontological commitment between the two alternatives, the acceptance of the explanation that makes use of the interpretability strategy only commits to the existence of the mentioned non-empty strings, whereas the mathematical explanation commits one to much more, that is, to that which comes with Frege Arithmetic.

Chapter 5
Conclusion Part I

Abstract The conclusion of the first part briefly summarizes the main arguments.

The main question Part I set out to answer was what the limitations internal to the EIA are, as an account for mathematical realism. Specifically, to answer the question under what conditions the argument can escape circularity. Given that the EIA relies on an IBE, the explanandum needs to be assumed as true for the argument to get off the ground. For this reason, it must not contain any references to mathematics. Paraphrasing the references to numbers in the explanandum into first-order logic with identity seems a viable strategy to get rid of the mathematical.

However, as a result of this strategy, it was observed that a gap between the explanandum and explanans emerges. The explanandum now contains a pluraliies, namely the collections of objects expressed by the sentence in first-order logic with identity, whereas the explanans is about an entities, namely the numbers. In other words, the context between explanandum and explanans differs. This is problematic, because how can primality be predicated of a collection of objects? After all, primality prima facie applies to numbers.

To mend this issue, some sort of bridge principle is needed, in order to connect entities and pluralities. To this end, HP was suggested, because it is a principle that precisely does that. However, it turns out that whether the EIA can escape circularity depends on the outcome of a debate on the status of HP. That is, it depends on the outcome whether the principle is analytical or not. For if HP turns out to be an analytical principle, then the EIA remains circular.

There is this bifurcation, if you like, quite early on. Or rather, hidden very deep in the argument. And it appears that the possibility of the argument stand or falls by the outcome of this different discussion. A discussion that seems far removed from the discussions regarding the EIA. But in fact, any example using number theory, and that needs to use the suggested paraphrasing strategy, will hinge on the outcome of the discussion of HP.

If it turns out that HP is not an analytical principle, then the EIA can escape circularity *in that respect*. Because even in the case HP is a non-analytical principle, the worry of the argument being circular still looms. Given the availability of a

© The Author(s), under exclusive license to Springer Nature Switzerland AG 2025
L. A. Tump, *On Mathematical Explanations of Empirical Phenomena*,
Synthese Library 517, https://doi.org/10.1007/978-3-031-98860-8_5

relative interpretation of the relevant mathematics of the explanation in a theory of concatenation (via Robinson's Q), then, by IBE, it still needs to be established that the mathematical explanation is the best explanation.

This also has to be done non-circularly. Specifically, one cannot appeal to mathematical strength (as discussed in Sect. 3.1), because that would be question begging for a nominalist. But if there were a difference in empirical scope between the explanation that makes use of the interpretability strategy and the mathematical explanation, then this would be unproblematic.

Of course, the question of the logical structure of the argument and whether it can escape circularity is hardly the last word about it, because then there is all the other literature with its criticisms (and defences) of it, some of which we have seen earlier. Regardless of the question whether mathematics does indeed do (part of) the explanatory work in the discussed example, it is clear that the debate on the status of so-called *abstraction principles*, such as HP, and work on relative interpretations of mathematical theories should be of great interest for anyone working on the topics of mathematical application and mathematical indispensability.

Part II
Mathematical Counterfactual Dependence

Chapter 6
Counterfactual Explanation and Mathematics

Abstract The second part shifts focus towards embedding mathematical explanations within a framework of counterfactual reasoning. This chapter sets up the context. It discusses a proposal for a non-interventionist counterfactual account for explanation. If the counterfactual information can indeed be separated from the causal information, then mathematical explanations could in principle be incorporated into a counterfactual theory of explanation. Because this inevitably leads to impossible antecedents, as mathematical properties are necessary, the topic of counterpossibles is introduced, specifically with regard to the question whether they are trivially true.

Part II of this book also focuses on the example of an extra-mathematical explanation introduced in Part I, but now in the context of a counterfactual theory of explanation. The idea here is to bring out the explanatory nature of mathematics counterfactually. As we have seen in the introduction, counterfactuals play an important role in accounts of causality and explanation. For obvious reasons, the causal accounts are not applicable to mathematics. However, the idea to bring out the explanatory relationship between a range of variables that an explanans can have counterfactually is applicable to mathematics. In this chapter, Sect. 6.1 discusses a proposal to adjust the counterfactual theory of explanation in such a way so as to disentangle the causal from the counterfactual.

Section 6.2 then provides the background relevant for the evaluation of the proposed example in the subsequent chapters, by discussing the topic of counterpossibles. This will be of use later on, since the counterfactual that is introduced in Chap. 7 is one that has an impossible antecedent. Thus, even though the mathematical properties are necessary, the idea is to try to vary these properties nonetheless, with the goal to bring out the explanatory contribution of mathematics to explanations. It does so by capturing number theoretic explanatoriness of the empirical phenomenon by counterfactually allowing for a change in the necessary property of the numbers in question.

In light of the example introduced in Part I, this means that the necessary property of the numbers representing the length of the life cycles of the Periodical Cicadas

© The Author(s), under exclusive license to Springer Nature Switzerland AG 2025
L. A. Tump, *On Mathematical Explanations of Empirical Phenomena*,
Synthese Library 517, https://doi.org/10.1007/978-3-031-98860-8_6

in years being prime will need to be changed. One of the reasons why such a counterfactual approach is suggested, is that a better understanding of the present example, but also mathematical explanation in general, is required. Specifically, in the context of the debate between the mathematical nominalist and mathematical platonist, an independently motivated understanding of mathematical explanation is required (Colyvan 2013, pp. 1043–1044). This is not to say that a choice for a theory of explanation is necessarily free of prior reasons for adhering to it, but it could nonetheless be progress, and less prone to intuition trading.

As mentioned in the introduction, the main question of Part II is what the effects of changing a mathematical antecedent of a counterfactual are. This is important, given that it is supposed to be part of the explanation of an empirical phenomenon. After all, in any explanation of an empirical phenomenon, the explanandum is a given. Given the resulting number-theoretic counterpossible, and keeping to our excercise of looking at the internal workings of the arguments, this amounts to the question whether we can be certain that we are still speaking about the same numbers in both the factual and the counterfactual case.

A recurring theme in these discussions is the strength of explanations. If there is more than one possible explanation of some empirical phenomenon, then looking at the degree of scope generality of the various possible explanations, with all the caveats mentioned in Part I (for instance in Sect. 3.1.1), could potentially guide us to the explanation that could be considered to be a better explanation in certain relevant respects. Mathematics, typically, has scope over a very large, if not infinite, domain of possible explanations, and therefore explanations that make use of mathematics have substantial counterfactual strength. And if it could be brought out that the explanation hinges on the mathematics invoked, this could be considered to be a good reason to believe those explanations have to be considered to be better than their counterparts with a more limited scope. However, as will be discussed, such an argument should be made non-circularly. That is, invoking the larger scope should not surpass the empirical context to which the explanations belong, on pain of begging the quesiton against the nominalist.

Section 6.3 will summarize the chapter and state the relevant aspects in light of the chapters to come.

6.1 Embedding Mathematical Explanation

As we have seen in Sect. 1.1, the basic idea of a counterfactual theory of explanation is that some pattern of counterfactual dependence explains how an explanandum would have been different, given some varying in the explanans. This provides us with what Woodward (2003, p. 11) calls 'what-if-things-had-been-different' information. The explanation thus shows the effect a change of some relevant aspects in the explanans has on the possible outcomes of the explanandum.

Essentially, on such a view, the counterfactual truths are seen as providing the basis for explanations (Saatsi 2016, p. 1061). Now, Woodward's initial counterfactual account of explanation is an interventionist account, that is, it is an account about what would happen given a certain intervention in relation to some phenomenon. In the example of the flagpole presented in Sect. 1.1, this could be the changing of the value of the variable relating to the height of the flagpole, and then look ate what the effect would be on some aspect of the explanandum in light of this intervention, i.e. the lenght of the shadow cast by the flagpole.

In Sect. 1.1 we have also seen that an intervention is inter-defined with causation. An intervention variable, I, causes a variable X to take a different value, and this value in turn has an effect on some other variable Y (Woodward 2003, p. 98). As mentioned before, the notion of causation has no place in mathematics. After all, causation is generally thought of as a relation between two events, and events are situated in space and time. Since mathematics is not situated in space and time, the notion of causation does not apply. If a counterfactual theory of explanation is to be applied to mathematical statements, then some changes to that theory are needed. In the first place, mathematical explanations are explanations that do not exhibit causal counterfactual dependencies, although they might exhibit non-causal counterfactual dependencies. In the second place, applying a counterfactual theory of explanation to mathematics requires for the necessary nature of mathematics to be accommodated.

With regard to the first aspect, one suggestion for the disentangling of counterfactuals and interventions is given by Reutlinger (2016). The general idea is that talk about causality should be separated from the general thesis of the counterfactual theory of explanation. Instead, the counterfactual theory of explanation will be formulated in terms of variations in the explanans. Let us follow Reutlinger (2016, p. 738) here, and say that E is the explanandum, and that $S_1, \ldots, S_n$ are statements of the explanans. Then, when the statements $S_1, \ldots, S_n$ would have been different, the explanans E would have been different as well. Reutlinger (2016, p. 737) calls this the 'dependency condition'. He considers something an explanation if and only if this dependency condition is met.[1] There is also a so-called 'possibility constraint' in relation to explanations. This means that the changes in the statements $S_1, \ldots, S_n$ should be possible given some generalization, such as a law, involved in the explanation.

The statements, $S_1, \ldots, S_n$, in the explanation could be specifications of initial conditions (similar for instance to those mentioned by Hempel and Oppenheim (1948) in their DN-theory of explanation), boundary conditions, bridge principles, or other assumptions. These other assumptions mostly tend to be about the how the models that are employed in the explanation are related. The generalizations

[1] Note that this is not a sufficient criterion. Reutlinger (2016, p. 737) also formulates what he calls a 'veridicality condition', i.e. the generalization and other statements in the explanation are (approximately) true, and an 'implication condition', i.e. the statement in the explanandum is logically entailed by the generalization and other statements in the explanation.

are then indeed the laws, or law-like statements, that are invoked in the explanation (Reutlinger 2016, pp. 736–737).

As such, the salient point of the analogy between Reutlinger's suggested non-interventionist account and Woodward's interventionist account is that both require a change in the value of some variable in the explanans. The difference is that, where the interventionist account requires the possibility of an intervention (on an intervention variable), the non-interventionist variety requires the possibility of at least one of the (values of the variables in the) statements of the explanans being of a different value. And where the intervention should, from I via X to Y, lead to a change in the value of explanandum, the change in value in the non-interventionist account should also lead to a change in the value of some variable in the explanandum, assuming the change fits the generalization(s) employed in the explanation (Reutlinger 2016, p. 738).

To bring out the difference between the interventionist counterfactual account and the non-interventionist counterfactual account of explanation, the example of Euler's explanation of the impossibility of a path through the original seven bridges of Königsberg, while crossing each bridge only once, is often used. In answering the why-question, i.e. why it is impossible to find the just mentioned specific path, no causal references are used. Rather, the explanation makes use of graph theory. What is relevant for the explanation, is the fact that there is no mapping between the actual lay out of the bridges of Königsberg (in the year 1736, that is) and a graph satisfying the necessary conditions for this kind of crossing of the bridges to be possible. This is due to what is now called 'Euler's theorem'. The theorem states that there is a Euler path through a graph if and only if it is connected (meaning that there is a path between any of the distinct vertices), and there is a path from an initial vertex using each edge only once in order to get back to it. For such a path to exist, Euler showed that each vertex must be of even degree, i.e. have an even number of edges (in casu, the bridges) (Pincock 2012, p. 51). In the explanation, the generalization being used is Euler's theorem. The contingent initial conditions, such as the number of bridges or the number of landmasses, are what can be varied, e.g. one could use vertices of an odd degree instead of an even, or edges of an even degree instead of odd. But none of these changes to the values of the initial variables have any causal bearing on the explanation.

We can now see that the idea that the (variables in the) statements in the explanans can take a different value is not very dissimilar to the idea of varying the object in a proof-idea in Steiner's (1978a) theory of mathematical explanation. For that reason, this idea of varying the object of a proof is of particular interest in relation to a counterfactual theory of explanation such as Woodward's (2003) or Reutlinger's (2016), that is, if non-causal explanations are to be accommodated by it.

One first step is to recognize the similarities in the role the employed generalizations have in these theories of mathematical explanation on the one hand, and theories of scientific explanation on the other. Gijsbers (2017, p. 54) points this out succinctly: where the generalization in the case of the mathematical model of explanation would be a proof-idea,[2] in the case of the interventionist model of explanation this would be a causal generalization, such as a natural law.

Taking Strevens's (2006) example of the baseball hitting the window again (see Sect. 1.1), then it is the causal generalization that links the force of the ball with which it hits the window (or the speed of the ball with which it hits it), and the subsequent breaking of the window. Taking Steiner's (1978a) example of the explanatory proof of the sum of the series again as well (see Sect. 1.2), then the proof-idea links the symmetry property of the constructed series with the sum of that series (Gijsbers 2017, p. 54).

The picture that arises is one of contrasting. The generalizations are kept fixed, and subsequently the variables pertaining to these generalizations are varied. By contrasting the different proofs emerging from the varying of the object within the proof-idea, we gain insight in the workings of some specific proof and why it holds. In the example provided in Sect. 1.2 on page 9, the sequence was varied, but the proof method itself was kept the same. Keeping the proof method the same resulted in different sequences giving different theorems. Thus, the way the proof gives us insight is by making use of the property of the symmetry of the sum of the natural numbers. Similarly, by contrasting the different possible situations the ball could be in when hitting (or not hitting) the window, we gain insight in the causal relations between the ball and the window in the specific case where the ball shatters the window. As such, the suggested modification of the counterfactual theory of explanation, in order to be able to accommodate non-causal explanations, can be thought of as replacing talk of interventions (on intervention variables) with talk of the varying of objects (within proof ideas).

However, recognizing the similarities between the two approaches is in itself of course not sufficient. For example, it appears to still be an open question whether all counterfactual information is in fact explanatory. Take the earlier mentioned problem of symmetry again, such as used in some counterexamples to the DN-theory of explanation. When counterfactual information does not require a causal dependency, then, when confronted with an equation in which variables can be tweaked to accommodate different scenarios, the different mathematically equivalent equations would all have to be considered to be explanatory (Saatsi & Pexton 2013, p. 618). For example, when the period of a pendulum is explained by the length of the rod, then, in virtue of the general 'law' as described by the formula

$$T = 2\pi\sqrt{l/g}$$

[2] Gijsbers (2017) calls this a 'proof-recipe', but the notion is essentially the same as Steiner's (1978a) proof-idea. The proof-recipe is rather a more elaborate version of this notion, because Steiner (1978a) did not work out the notion of a proof-idea. Additionally, the idea of a proof-recipe also builds on improvements from Weber and Verhoeven (2002), specifically with regard to the idea that for a proof to be explanatory, it in fact exceeds a single proof.

with T being the period, l being the length of the rod, and g the gravitational field,[3] the period would also explain the length of the rod of the pendulum:

$$l = T^2 g / 4\pi^2$$

Here, the causal aspect appears to be fundamental for determining the explanatory direction of the equation in this particular example. Or rather, as Saatsi and Pexton (2013) put it, in the specific instance of the generality invoked in this specific example.

Saatsi and Pexton (2013) also suggest that the causal aspect and the counterfactual aspect can be disentangled, but that one of the consequences of such a separation would be that some form of unity between explanations might need to be given up (Saatsi & Pexton 2013, p. 615). To substantiate this claim, they point to a distinction between explanations of single states of affair, such as a specific pendulum with a certain length, on the one hand, and what they call 'regularity explanations' on the other. The latter kind of explanation emerges when making the general statement that, for example, the length l of some pendulum is proportional to the strength of some gravitational field g (Saatsi & Pexton 2013, p. 622).

What is important with regard to this distinction is that, when considering counterfactual information, the causal dependencies, such as the period of the pendulum's dependence on the length of the rod, ensure what bits of counterfactual information can be considered as explanatory *in single states of affair*. What Woodward has shown, they say, is that causality plays an indispensable role in determining the asymmetry in such specific cases, but not in that of the more general regularity explanations (Saatsi & Pexton 2013, p. 621). With regard to single states of affair, it would be so that if the intervention would not have occurred, the effect of it would not have occurred either. Stating the idea this way is indeed very similar to Hume's statement already referred to in Sect. 1.1, and from which Lewis took cue for developing his counterfactual theory of causation: "if the first object had not been, the second never had existed" (Lewis 1973a, p. 556). However, when it comes to regularity expressions things are different, for it might well be the case that a certain effect E obtains precisely when an intervention I is performed. But of course, for something to be an explanation, the co-occurrence of E and I is not a sufficient condition. We also need to know what would have happened in the case where the intervention had not taken place. Therefore, contrary to regularity explanations, modal information is often not sufficient for the explanation of single states of affair, and hence the often indispensable use of the notion of causation in such cases.

What is more, Saatsi and Pexton (2013, p. 622) also argue that in the case of regularity explanations, intuitions about the asymmetry of explanations are much weaker or even non-existent. In regularity explanations, directionality seems much

[3] For simplicity, the simple formula abstracts away from other aspects that influence the period of the pendulum, such as its mass or the resistance of the air.

less of a problem and the one side of the regularity can be used to explain the other just as well as the other side can be used to explain the one. To see this, let us go back to the above mentioned general 'law' describing the relation bewteen the length of a rod and its period. The observation is that there is no prima facie problem with considering the derivation of the fact that lenghts of rods are proportional to their periods, or even to the strength of the gravitational field, as explanatory (p. 622). The reason for this is that the explanation hinges on counterfactual information rather than causal information, and the possible outcomes are all situated in the, to use Steiner's terminology, proof-idea. This indeed requires an abandonment of the interventionist assumption.

In any case, there appears to be no prima facie problem with explaining a regularity by citing the relevant modal information for either direction.[4] In fact, when empirical laws are concerned, this appears to be quite often the case. Hempel (1962, p. 12) already made a similar observation in this regard, when he stated that not all DN-explanations are causal explanations. For example, he stated that this would be the case when a law is subsumed under a more general law, or when universal facts are concerned.[5] The important takeaway from the above observations is that counterfactual information is often seen as coming first with regard to explanations. In many instances this counterfactual information can, but only secondary, be causally interpreted (Saatsi 2016, p. 1061).

The second difference between the non-interventionist and interventionist interpretation of the counterfactual account, mentioned at the beginning of this section, is that in the case of the interventionist interpretation some property is varied, for instance the speed of the ball, or the mass of the ball. This is possible due to the contingent nature of physical objects. After all, the ball could have traveled at a different speed, or have had a different mass.

With regard to mathematics, it is not immediately clear what is meant when speaking of varying the property of a mathematical object. Due to the necessary nature of mathematics, mathematical objects cannot but have the properties they have. Their properties could not have been different. Of course, in some sense mathematical objects can have contingent properties. For example with regard to numbers, when some specific number is someone's favourite number. But this is quite different from that number having the property of being, say, prime. With regard to this issue, Gijsbers (2017, p. 54) gives the example of the symmetry property in Steiner's (1978a) example of an explanatory proof. The series in the example cannot have another symmetry property than the symmetry property it actually has. In other words, the property cannot be modified or adjusted. In fact,

[4] It is interesting to see that in this case the modal aspect seems to be parasitic on the logical implication in the regularity.

[5] In fact, Hempel goes on to say that there are explanations of particular facts which cannot be considered to be causal, such as instances of Boyle's law, which law states that the pressure of a gas at a specific time and of an unchanged temperature is inversely proportional to its volume. As should be clear form the previous sentence, this is also an example of a case where proportionality is concerned (Hempel 1962, p. 12).

it is for this reason that Steiner talks about varying the object itself, rather than the property. As a result of the necessary nature of mathematics, the structure, or domain, within which the object is varied is kept the same when substituting the one object for the other object. In Steiner's example this means substituting the one sequence of numbers for the other sequence of numbers.

Despite these observations, in the next Chap. 7, we will nonetheless have a look at a proposal in which the idea precisely is to 'change' a mathematical property. Doing so inevitably leads to a counterpossible statement, that is, a statement with an impossible antecedent, due to the necessary nature of mathematics. In the example the property that is varied is the primality of the numbers pertaining to the lenght in years of the life cycles of the Periodical Cicada. The idea is to bring out the explanatory contribution of mathematics in relation to this empirical phenomenon.

6.2 Counterpossibles

Before moving on to the analysis of the suggested example of an extra-mathematical explanation in which some necessary properties of mathematics are counterfactually 'changed' to bring out its explanatory role, a short digression on the topic of counterpossibles itself is in place. The analysis of the proposed example is after all embedded in the broader discussions on this topic. In particular, the question of whether counterpossibles are vacuously true, as the standard account endorses, is of importance in the current context. The reason for this is that, if it is indeed the case that counterpossibles are vacuously true, then doubt is cast upon the possibility for there being any extra-mathematical explanatory role for such statements. The accompanying idea here is that vacuously true statements cannot be explanatory, because when the truth of a subjunctive statement, if it is intended to be (part of) an explanation is trivialized, then it does not make a difference to that explanation. And of something that is supposed to contribute to an explanation, it is expected to make a difference to it. Although we will not take a position in the debate itself here, some of the difficulties highlighted in this section are relevant for the further discussions on the topic in the subsequent Chaps. 7 and 8.

Generally speaking, counterpossibles are a variety of counterfactuals where, contrary to ordinary counterfactuals, that which is expressed in the antecedent is necessarily false. As such, in the case of a counterpossible, the 'what-if-things-had-been-different' question appears to be somewhat strange. At least, in the sense that the thing that is supposed to be different could not have been different in the first place, as it is an impossibility. The question is then how to evaluate such statements?

The question as to the evaluation of counterpossibles is particularly relevant for our discussion, as contributions to explanations are generally thought of as making a difference to the explanation. However, on the standard semantics for counterfactuals, as it was initially developed by Stalnaker (1968) and Lewis (1973b), counterpossibles are considered to be trivially true. This is a direct result of the necessarily false antecedent, and the evaluation of a counterfactual statement depends on the truth value of its components, i.e. the antecedent and consequent, in some alternative situation to the factual one. Take as an example the following counterfactual by Lewis (1973b, p. 1):

(1) if kangaroos had no tails, they would topple over.

If we turn to the standard semantics for counterfactuals for the evaluation of the above statement, then we have a look at those situations, or possible worlds, where kangaroos have no tails and check if these kangaroos topple over. If in all the worlds where kangaroos have no tails, they topple over, then (1) evaluates to true. And, if in some world where kangaroos have no tails and they do not topple over, then (1) evaluates to false. To put it more generally, a counterfactual $A \mathrel{\Box\!\!\rightarrow} B$ is said to be true at a world w if and only if it is the case that the closest A-worlds are B-worlds. In other words, there is no A-world that at the same time is not a B-world that is closer to world w (Walters 2016, p. 323).

This evaluation already implies that not all possible worlds w are relevant for evaluation. Only those A-worlds that are similar in the relevant respect to the actual world are, or similar in the relevant respect to the possible world in which the counterfactual is stated, are up for consideration. In the case of the tailless kangaroos, for instance, only worlds with kangaroos are relevant. Worlds in which no kangaroos exist are irrelevant for the evaluation of the counterfactual. The more similar the possible worlds are to the actual world, the closer to the actual world they are said to be. To accomodate closeness, the relation between the different worlds has to be such that the related worlds are indeed sufficiently similar in the relevant sense. This similarity is contextually determined (Berto & Jago 2019, p. 20). Of course, what 'sufficiently similar' amounts to is a discussion in and of itself, but what seems to be clear is that the concept of *similarity* harbours some inherent vagueness. This, however, need not bother us here.

Thus, returning to counterpossible statements, we see that when their truth conditions are evaluated using the standard semantics for counterfactuals, then these kind of statements trivially come out as true, the reason for this residing in the fact that the set of worlds in which A occurs in the case of a counterpossible is the empty set, in turn due to the impossibility of the stated change of the property. Take for instance the sentences (Lewis 1973b, p. 25):

(2) if there were a largest prime p, $p! + 1$ would be prime.[6]
(3) if there were a largest prime p, $p! + 1$ would be composite.

Because of the infinite sequence of numbers, no possible world can contain a largest prime number, and hence both of the sentences come out as true on the standard semantics. That is, they are vacuously true. There does not exist any A-world, and by consequence there is no closest A-world either. Therefore, the universally quantified statement of the form 'for all closest A-worlds ... ', is trivially true.

---

[6] The factorial of a natural number, denoted as $n!$, is the product of all numbers in the range $[1 \ldots n]$.

The evaluations of above sentences (2) and (3), and the more general point that all counterpossibles are vacuously true as well, is accepted and defended by someone like Williamson (2018). In these particular cases, for example, it is accepted because it is consonant with valid mathematical reasoning. Effectively, the sentences summarize a proof by *reductio at absurdum* that there is no largest prime number. The proof itself makes use of the fundamental theorem of arithmetic, which states that every positive integer has a unique prime factorization, or is itself prime. In short, the proof is as follows: suppose that p is the largest prime number. Then $p!$ would be divisible by all prime numbers, because $p!$ is divisible by all the positive integers. From this, it follows that $p! + 1$ is not divisible by any (prime) number, and hence $p! + 1$ would be prime (2). But at the same time, and contradictory, it follows that resulting from $p! + 1$ being larger than the largest prime p, $p! + 1$ must be composite. Taken together, (2) and (3) lead to the following contradiction (Williamson 2018, p. 363):

(4) if there were a largest prime p, $p! + 1$ would be both prime and composite.

Based on this contradiction, the conclusion that there is indeed no largest prime number is then justified.

What is generally considered to be the standard argument for the vacuous truth of counterpossibles is provided by Williamson. See for instance (Williamson 2018, p. 358), and goes back to Lewis and Stalnaker. The argument can be stated that given the fact that a strict implication implies a counterfactual, $\Box(\alpha \rightarrow \beta) \rightarrow (\alpha \mathrel{\Box\!\!\rightarrow} \beta)$, then, by the further fact that an impossibility implies everything in standard modal logic, it also follows that $\Box\neg\alpha \rightarrow (\alpha \rightarrow \beta)$, and therefore by transitivity $\Box\neg\alpha \rightarrow (\alpha \mathrel{\Box\!\!\rightarrow} \beta)$.

However, not everyone agrees with the idea that all counterpossibles are trivially true. Some think that there exist counterpossibles that might be false or non-trivially true. A classic example from the literature to illustrate the idea behind this thought, is captured in the following sentence (Nolan 1997, p. 544):

(5) if Hobbes had (secretly) squared the circle, sick children in the mountains of South America at the time would have cared.

With regard to this example, the intuition some have is that (5) is false, rather than vacuously true. Or, when 'would have cared' were to be substituted by 'would not have cared' the sentence would be true non-vacuously (Brogaard & Salerno 2013) and (Berto et al. 2018). This intuition comes from the idea that the sick children would indeed not care at all, probably because they have more pressing things on their minds. And that therefore, the counterpossible (5) is false.

In response, Williamson (2018, p. 364) observes that, when reasoning along these lines, the antecedent appears to play no role at all in the evaluation of the truth value of the sentence. The squaring of the circle could have been replaced by anything else, for that matter. According to him, what happens is that a pre-theoretic thought process kicks in when comparing the two options regarding the opinions of the children, i.e. caring or not caring about the squaring of the circle by Hobbes. And because it is believed, exempli gratia, that the sick children do not care, a sentence in which the children do not care is considered to be true. As a result of this, there is a strong inclination to consider a sentence like (5), in which in contrast the children do care, to be false. But when theory comes in, Williamson continues, then we *do* take into account the antecedent in the evaluation of the counterpossible. And so, on the standard account, (5) would also come out as true.

In any case, if the intuitions of the non-vacuists turn out to be correct, then it would not make the counterpossible vacuously true. And as a consequence, it would turn out that the standard semantics for counterfactuals would not tell the whole story about counterpossibles. Much of the debate on the truth value of counterpossibles is however focused on counterexamples and intuitions about those, and examples might be somewhat unfortunate in some way or other.

In order to avoid a deadlock in intuition trading, and to solely work with examples tailored for specific philosophical purposes, one strand in the debate looks to other domains than philosophy for counterexamples to the standard account. The idea is to look for examples of counterpossibles that seem to be treated as non-vacuously true (or as false) in these other domains. The idea is that if such examples can indeed be found, then this would constitute good reasons for the standard semantics for counterfactuals to have to give way. Take for instance the following two examples by Jenny (2018, p. 530):[7]

(6) if the validity problem were algorithmically decidable, then the halting problem would also be algorithmically decidable.
(7) if the validity problem were algorithmically decidable, then arithmetical truth would also be algorithmically decidable.

In the field of relative computability theory, counterpossible (6) appears to be considered to be true. The reason for this is that the halting problem is Turing reducible to the validity problem (Boolos et al. 2007, pp. 126–132). Counterpossible (7), on the other hand, appears to be considered to be false, because arithmetical truth is not Turing reducible to the validity problem.

Reduction, or rather Turing reduction, from one problem to another problem basically means that if there is a method to algorithmically decide the one problem (for instance the validity problem), then this method could be used to decide (instances of) the other problem (for instance the halting problem). To give a simple example, if we want to solve problems of multiplication, we could reduce those to problems of addition.

However, it could also be argued that such counterpossibles are in fact undergirded by proofs. Williamson (2018, pp. 361–362), for example, states that a claim used in a counterpossible, for instance a claim such as expressed in (6) and (7), must have an epistemically relevant property, for example a proof, for it to be of use. In

[7] The validity problem, or decision problem, states that first-order logic is undecidable. There is no method for deciding whether a finite set of (number encoded) sentences are logically valid; the halting problem is about deciding whether a computer program will stop running (will halt) given arbitrary input; the decidability of arithmetical truth states that a system cannot give the truth value for all of its sentences (Jenny 2018, p. 532).

that sense, the truth of the counterpossible is actually a consequence of the proof. The question would then be what the relation between proof and truth is. Specifically in relation to assertability conditions. The truth of a statement alone appears not to be sufficient for asserting that statement in a proof.

The important observation for us is that the debate on the status of counterpossibles is an ongoing one, and, as mentioned before, no position will be taken in this debate here. However, it is the case that the outcome of this debate, or the position one takes, is important for the evaluation of the counterpossible that will be introduced in Chap. 7, because it is meant to play a role in an explanation of an empirical phenomenon. Therefore, as previously already discussed in Sect. 4.1, the statement used in the explanans needs to be true. Or at least assumed to be true. More so, in the context of extra-mathematical explanations, the counterfactual has to be non-trivially true because everything that is trivially true is not essential in an explanation. As we have already seen, Williamson (2018), for example, defend the view that counterpossibles are trivially, or vacuously, true. But others, such as Priest (2016) or Berto et al. (2018), defend the idea that counterpossibles can be non-vacuously true.

Given that the tweaking of the mathematics in the antecedent leads to inconsistencies, then, on the standard semantics for counterfactuals, the consequent has no role to play in the evaluation of the truth value of the counterpossible. In other words, whether the consequent is B or $\neg B$, or whatever it might contain, in both cases the counterpossible evaluates to true. If the mathematics, and specifically that part of the mathematics that is changed, is to be brought out as carrying explanatory load, then it has to make a difference to the explanation, as is expected of difference makers in general. But when trivialized, this role cannot be brought out (Baron 2016, p. 380). Therefore, it must be possible for there to be non-vacuously true (or false) counterpossibles. A similar observation is made in the literature on the logic of so-called 'why-questions', which is intimately related to the topic of explanation. For something to be an answer to a why-question it needs to be non-trivially deducible from the explanans (Hintikka & Halonen 1999, p. 195). For these reasons, the outcome of the status of counterpossibles is relevant for the evaluation of the example discussed in the subsequent chapters.

6.3 Conclusion

Section 6.1 of this chapter started out by discussing a proposal for a non-interventionist counterfactual account for explanation. If the counterfactual information can indeed be separated from the causal information, then mathematical explanations could in principle be incorporated into a counterfactual theory of explanation. This is relevant for the topic under discussion in this book, because this would mean that there is an externally motivated theory of explanation available that does not prejudge the position of the mathematical nominalist or platonist. The salient observation is that the counterfactual information is seen as the main source for explanations, which in many instances can be interpreted causally, but not necesarily in all.

For the mathematical cases, the idea of the varying of the mathematical object was taken from Steiner's theory of mathematical explanation. This resulted from the observation that mathematical properties are necessary, from which it is concluded that its properties cannot be varied.

Regarding the necessary nature of mathematics, Sect. 6.2 introduced the topic of counterpossibles. This is immediately relevant for the example of an extra-mathematical explanation within the context of a counterfactual theory of explanation introduced in Chap. 7. Given the nature of the example, namely that it contains an impossible antecedent precisely because it intends to vary a mathematical property, the discussion on counterpossibles is of importance. Specifically, since the counterpossible is part of a deductive explanation, it has to be non-trivially true, because if it were trivially true, it would not be essential in the explanation. The outcome of this debate, if at all a consensus is reached, will play a role in the evaluation of the suggested example later on. What is more, because the non-trivial truth of the counterpossible is needed, the relevance of the connection between the antecedent and consequent cannot be ignored either.

Chapter 7
A Number-Theoretic Counterpossible

Abstract This chapter discusses a concrete example of counterfactual reasoning about a mathematical explanation of physical facts, which serves a starting point for number-theoretic counterpossibles. Special attention will be given to contradictions that arise from varying the necessary properties in the antecedent. Even though a contradiction tolerant logic would be able to deal with those, this is not the only problem. Since the numbers are not closed under succession anymore, the resulting model of arithmetic will not be isomorphic to the standard interpretation.

In the philosophy of explanation, counterfactual reasoning is not often geared towards extra-mathematical explanations. Although for instance, as mentioned in Sect. 3.1.1, Knowles and Saatsi (2019) provide an analysis of the role mathematics plays in relation to the so-called 'explanatory generality' of extra-mathematical explanations. In this context, generality pertains to the ability of mathematics to connect seemingly unrelated empirical phenomena under a single schema. The ability to do so, is often considered to be one of the key virtues of the explanatory role of mathematics. The analysis they carried out was also done out within the framework of a counterfactual theory of explanation. One of the reasons why this analysis was carried out within the framework of a counterfactual theory of explanation, was to have an externally motivated theory which should provide a relatively neutral starting point.

In their (2017) article, Baron et al. provide an account of counterfactual reasoning about mathematical explanations of physical facts. Their aim is to extend the counterfactual theory of explanation in such a way, so as to be able to include those. We remember that extra-mathematical explanations are explanations in which the mathematics contributes to the scientific explanations themselves. The account, then, would also purportedly open the door for an argument for mathematical platonism. Additionally, in their (2020) article, they also expand on their account to include intra-mathematical explanations, that is, explanations of mathematical facts.

The general idea of Baron, Colyvan, and Ripley is similar to the basic idea of the counterfactual account of explanation, or more generally, counterfactual reasoning. They intend to show that some pattern of counterfactual dependence explains how

© The Author(s), under exclusive license to Springer Nature Switzerland AG 2025
L. A. Tump, *On Mathematical Explanations of Empirical Phenomena*,
Synthese Library 517, https://doi.org/10.1007/978-3-031-98860-8_7

an explanandum would have been different given a different explanans. As in other well-known examples of counterfactual reasoning, such as in 'if Suzy had not thrown the rock, the bottle would not have shattered', some parts of the facts are kept fixed. Subsequently, the antecedent is varied, and the implications of the varied antecedent are then considered. Or, in the terms they use, some facts are held fixed, some others are twiddled, and then one looks at the ramifications of the twiddling (Baron et al., 2017).

In the example of Suzy one might, as Lewis (1973b) suggested, keep the past up to the moment Suzy throws fixed, then vary the fact that Suzy threw the rock (i.e. that she didn't throw the rock in the counterfactual situation), and then look at the implications of the varying of the antecedent (i.e. that the bottle did not break in the counterfactual situation).

This general idea is then also applied to extra-mathematical explanations in which the counterfactuals have a mathematical antecedent. In such cases, the 'what-if-things-had-been-different question' stipulates something that is different in the mathematical antecedent of the extra-mathematical explanation, perhaps a change in a mathematical property, and subsequently one looks at the effect this change has on the explanandum. Specifically, in their examples, it is applied to the explanation of the prime life cycles of the North American Periodical Cicada and the explanation of the hexagonal shape of honeycomb cells (Baron et al., 2017). It needs to be emphasized that they explicitly assume there to be genuine cases of mathematical explanations of empirical phenomena, including the just mentioned explanations, and that their aim is not to discuss the specific examples in themselves, but rather to develop a theoretical understanding of such cases (Baron et al., 2017, p. 2). Since we are interested in the internal workings of such arguments, we will also put this issue aside for now.

The question with regard to such an account, is how to deal with counterfactual statements in which the antecedent has a mathematical component. In the specific example under discussion, the focus will be on counterfactuals with number-theoretic antecedents. The difficulty lies in the fact that tweaking the mathematics in the antecedent inevitably leads to impossibilities, given the necessary nature of mathematics. Due to the important role the context plays in determining the truth conditions of counterfactuals in general, and counterpossibles in particular, the question then is how to understand what holding the context fixed could mean whilst tweaking some part of the mathematics involved in the explanation, which, as we will see shortly, is done in the example pertaining to number theory.

Section 7.1 introduces the example of the mathematical counterfactual and points to some immediate worries concerning contradictions that might arise. Section 7.2 discusses the relevance of such contradictions and argues, by means of Peano Arithmetic, which is much used in meta-mathematics, that these contradictions point to an underlying difficulty. Section 7.3 will summarize the salient points and in Chap. 8 it will be argued that, as a result of the twiddles in the number-theoretic antecedent, in effect a shift in the structure of the natural numbers takes place. The arguments draw on discussions on the standard interpretation of arithmetic and de re knowledge of natural numbers. The importance of the idea of being isomorphic to the standard model of arithmetic and of the idea of succession, exemplified in the successor function of the Peano axioms, are discussed. Subsequently it is noticed that for a counterfactual to be the counterfactual of something or other, it needs to be *about* this something or other. The change of structure casts doubt on this. Additionally, and also resulting from the changes in the context antecedent, it will also be questioned whether it is possible for the empirical content of the consequent, which is a given, to latch on to the antecedent.

7.1 A Mathematical Counterfactual

Let us first briefly remember the case of the Periodical Cicada, as introduced in Chap. 2, and of which the life cycles are either 13 or 17 years. It is observed that these life cycles are prime, and number theory could potentially explain that, given certain biological and ecological constraints, prime life cycles minimize intersection with other life cycles. In turn, this minimizing of intersection with other life cycles could be advantageous because the life cycles or the Periodical Cicada are now less likely to intersect with the life cycles of, say, predators (see Baker, 2005). Whatever the biological status might be of the above explanation of the empirical phenomenon, what Baron, Colyvan, and Ripley are concerned with is the embedding of these kind of extra-mathematical explanations in the general framework of a counterfactual theory of explanation.

Therefore, they consider the following counterfactual statement about the length of the life cycles (Baron et al., 2017, p. 6):

(1) If [...] 13 [...] were not prime, then these Cicadas would not have had 13-year life-cycles lengths.

According to them, this would counterfactually capture the explanatory role played by the mathematics. Since the explanation of the length of the life cycles turns on facts about the number 13, namely that it is a prime number, and also on the number-theoretic theorem about primes and common multiples, namely that prime cycles intersect less with other cycles compared to non-prime cycles, therefore considering 13 to not be prime would remove the support of the explanation that a 13-year life cycle is an advantageous life cycle to have because of it intersecting less with other periodical life cycles (given certain ecological and biological constraints).

Here, it is instructive to notice that the above counterfactual is indeed different from a counterfactual such as:

(2) If the length of the life cycle were not prime, then these Cicadas would not have developed a life cycle of that length.

In the above counterfactual, the twiddling is not done on the mathematics, but on the length of the actual life cycles. The salient difference is that in (1) we are dealing with an impossible antecedent, because mathematical truths are considered to be necessary truths. After all, 13 is a prime number *necessarily*. In (2) we are however dealing with a possible antecedent, because the length of the life cycle is a contingent truth. There is nothing that necessitates the specific length, and, for all we know, it could have been different.

Baron et al. (2017, p. 6) then observe that in order to consider 13 as not being a prime number, it would presumably mean that it is composite. And therefore, that it would not only factorize in 1 and 13, but, in line with the properties of composite numbers, also in some other numbers. It is imagined that, since 13 is not supposed to be prime, it counterfactually also has 2 and 6 as factors. To be sure, nothing hangs on this particular choice. The point is that some other factorization must presumably be the case. Given these hypothetical factors, the counterfactual could then also be restated as follows:

(3) If, in addition to 13 and 1, 13 had the factors 2 and 6, North American Periodical Cicadas would not have 13-year life cycles.[1]

In line with the general idea of a counterfactual theory of explanation, some facts need to be kept fixed and some facts are then varied. In the case of the sentences expressed in (1) or (3), the idea of Baron, Colyvan, and Ripley is to

> hold all of number theory fixed except for the twiddles to 13 if one is prepared to change the way that multiplication works (Baron et al., 2017, p. 7).

They therefore stipulate a new function other than multiplication, called 'multiplication*', which

> will preserve the same theorems as multiplication, and imbue the natural numbers with the same structure, except for whatever disruption is involved in changing the factors of 13 (Baron et al., 2017, p. 8).

But if the idea is to only change the factors of the number 13 and keep the rest of number theory as it is, a contradiction would immediately follow. This is so, because, for instance, $2 \cdot 6 = 12$, but since 2 and 6 are now also factors of 13 in the stipulated function multiplication*, this means that $2 \cdot 6 = 13$ is also the case. But then at the same time $12 \neq 13$ also holds, and so a contradiction arises.

It looks like there is no way to continue, and that by reductio it has been shown that 13 cannot be not prime. Or, in order to accommodate the contradiction, either the principle of non-contradiction has to be abandoned, or one needs to resort to some paraconsistent logic. However, Baron et al. (2017, pp. 7–8) state that this need not necessarily be so. The suggestion is that, when at first 13 is varied (as also having, besides 1 and 13, factors 2 and 6), and a contradiction is obtained in a manner as just described, one relaxes some of what was held fixed in the mathematics before the initial twiddle was done, and then repeats the process of varying again. This then goes on until the maximum amount that can be held fixed is found without having a contradiction. Although not all of number theory

[1] Note that, strictly speaking, this restated counterfactual is more precise than the original counterfactual as expressed in (1). For the present purposes however, this is of no real importance.

is preserved this way, presumably enough is. This echoes a similar point made by Berto et al. (2018, p. 704):

> whether a particular chain of reasoning succeeds or fails in supporting the truth of a counterfactual depends on the context, and in particular what truths about the case need to be held fixed to legitimate the reasoning.

In a similar vein, Baron et al. (2017, p. 9) state that not all contradictions that arise need to be addressed, but only the relevant ones. The metaphor they use is that of ironing out a lump in the carpet away from our area of interest. The lump might still be somewhere toward the side of the carpet, but not in the area that we are interested in. This would suffice, they say, because in a similar manner in the case of Suzy counterfactually not throwing the ball, there are also bound to be some contradictions. Consider the following examples: "Suzy moved her arm in a throwing motion, yet the rock did not move? She willed her arm to move, but it didn't?" (Baron et al., 2017, p. 9)

However, it is important briefly reflect on this. One thing is that it is not immediately clear whether in those cases there are bound to be contradictions in the same sense as there are bound to be contradictions in, say, the case of the change in factors of the number 13. This has much to do with the difference between the contingent and the necessary. Take for instance the example of Suzy moving her arm in a throwing motion, but with the rock not moving. Surely the laws of physics are kept the same in this case, because this is relevant to the assessment of the counterfactual 'if Suzy had not thrown the rock, the bottle would not have shattered'. Otherwise nothing could be said about the bottle shattering at all. Or, in the case of her willing her arm to move, but the arm did not. Is that a contradiction? Prima facie, this does not seem to be the case. Another question is whether it is relevant for the assessment of the counterfactual itself? In fact, it could argued that neither of the examples are in fact real contradictions. They are not inconsistent with, for example, classical physics, as long as one is willing to incorporate, say, extra forces. Although perhaps uninteresting from the viewpoint of physical theory this might seem, logically speaking there is nothing to prevent one from doing so. To put it in the style of Lewis: although in the case of Suzy a miracle might have occurred, it was however no logical contradiction.

Putting such questions aside for now, we are left with the aspect of the relevance of the contradictions arising in relation to the mathematical counterfactual, and the subsequent question whether they need to be addressed at all. Baron et al. (2017, p. 9) state this as follows:

> In a way, our example of the contradiction generated by multiplication (once we suppose that 13 is composite) is a little misleading. After all, it is not clear that any of this is relevant to the counterpossible of interest, so there's no need to even consider the contradictions arising there. We presented the details of how we might iron out such a contradiction by way of example. But, in fact, there is no reason to think that such further contradictions are relevant.

Section 7.2 will address the question of the relevance of the contradictions in the case of the mathematical counterfactual at hand, and investigates whether there could be reasons to think these contradictions might be relevant after all.

7.2 The Relevance of the Contradiction

Let us first begin with some short remarks on the earlier-mentioned contradiction arising from the twiddling of the number 13, namely that $2 \cdot 6 = 12$ and $2 \cdot 6 = 13$, but that $12 \neq 13$. Much depends of course on how the function multiplication* is actually defined. Let us assume for now that the rest of number theory is kept the same, and that we only have to deal with this specific contradiction. In that were the case, we could probably continue from here as follows: given that, for instance, multiplication* factors 13 in 1, 2, 6, and 13, this would mean that $2 \cdot 6 = 13$. But then it might also mean that $4 \cdot 3 = 13$ because $(2 \cdot 2) \cdot (6 \div 2) = 13$, if division needs to be changed as a result as well. And so on. A worry that one might have is that a local twiddle reverberates through the whole of number theory after all (Baron et al., 2017, pp. 7–8).

Since it is not deemed necessary to immediately resort to a paraconsistent logic, something we will briefly address later on, a method is proposed to deal with the ironing out of the (relevant) contradictions. It goes as follows: if a contradiction arises due to the twiddling of the number 13, then some more of the mathematics needs to be varied. If another contradiction arises, then some more of the mathematics needs to be varied still, until one reaches a point without a (relevant) contradiction (Baron et al., 2017, pp. 7–8). This way, the relevant contradictions should at some point be contained. Apart from describing this general method, it is not really shown how, and if, this would work. How to contain the contradictions arising from the twiddling of the number 13 in the context of number theory? Remember that multiplication* needs to "imbue the natural numbers with the same structure, except for whatever disruption is involved in changing the factors of 13" (Baron et al., 2017, p. 8). This therefore raises the further question what other disruptions there might be when changing the factors of 13.

Since we are concerned with number theory, a good starting point to evaluate the possibility of the twiddles to multiplication, and the effects such a twiddle might have on other parts of number theory, is Peano Arithmetic. Peano arithmetic is after all meant to define the natural numbers non-circularly. Notions like 'number' or 'successor' are meant to be captured by the axioms.

The language of arithmetic is described as follows, $\mathcal{L}_A = (0, S, +, \cdot, =)$, with 0 a constant, S the function symbol for successor, $+$ and $\cdot$ the function symbols for addition and multiplication respectively, and $=$ the symbol for the equality relation. Peano arithmetic consists of the Peano axioms, which are intended to define the set of natural numbers $\mathbb{N}$, and the functions for addition and multiplication (cf. Van Dalen, 2004, p. 87):

Peano Axioms

1. $\forall x(0 \neq S(x))$
2. $\forall x \forall y((S(x) = S(y)) \rightarrow x = y)$
3. $\varphi(0) \wedge \forall x(\varphi(x) \rightarrow \varphi(S(x))) \rightarrow \forall x \varphi(x)$

Addition
4. $\forall x(x + 0 = x)$
5. $\forall x \forall y(x + S(y) = S(x + y))$

Multiplication
6. $\forall x(x \cdot 0 = 0)$
7. $\forall x \forall y(x \cdot S(y) = x + (x \cdot y))$

The first axioms states that there is an initial element, denoted by 0, and which is not the successor of any element. The second axiom states that different elements have different successors. The third axiom, or rather the axiom scheme known as the 'standard induction scheme', is the first-order implementation of the axiom of mathematical induction. We can see how the earlier mentioned notions such as 'number' or 'successor' are meant to be captured by the axioms. The other axioms for addition and multiplication are recursively defined and self-explanatory.

Given the Peano axioms, with the Peano numerals as the names for the numbers, it is clear why $\overline{2} \cdot \overline{6} = \overline{12}$ and $\overline{2} \cdot \overline{6} = \overline{13}$ give rise to a contradiction in the first place, i.e. that $\overline{12}$ cannot be equal to $\overline{13}$.[2] This is so because, generalizing the point, it is the case that the product of the two functions of multiplication and multiplication*, given some numbers m and n, will agree in their outcome eventually. And this violates the requirement of uniqueness, as per axiom 2. Let us be more precise and use $\cdot$ for multiplication, and $\cdot^*$ for multiplication*. Take as an example $m = \overline{6}$ and $n = \overline{2}$ (or vice versa). Then, given the recursive definition of multiplication (axiom 7), it can be deduced that the following holds for multiplication*:

$$\overline{13} = \overline{6} \cdot^* S(\overline{1}) = \overline{6} + (\overline{6} \cdot^* \overline{1}) = \overline{6} + (\overline{6} \cdot \overline{1}) = \overline{12}$$

Considered this way, $\overline{13} = \overline{12}$, unless it would be the case that $\overline{6} \cdot^* S(\overline{1}) \neq \overline{6} + (\overline{6} \cdot^* \overline{1})$.

We now enter the terrain of the question of the containment of the contradiction. For the function of multiplication* to be able to deal with this initial contradiction, a first approach could be to define it in an 'exceptionalist' way. For instance, by taking multiplication for all m, n other than $\overline{2}$ and $\overline{6}$, and then define $\overline{2}$ and $\overline{6}$ separately, such that $\overline{2} \cdot^* \overline{6} = \overline{13} = \overline{6} \cdot^* \overline{2}$. Accommodating this in the new function of multiplication* means that the recursive step in the axiom of multiplication would then also have to be updated and replaced by the following:

$$\forall x \forall y(\neg((x = \overline{6} \wedge y = S(\overline{1})) \vee (x = \overline{2} \wedge y = S(\overline{5}))) \rightarrow (x \cdot S(y) = (x + (x \cdot y))))$$

However, replacing this step also means that $\overline{6} \cdot S(\overline{1}) = \overline{6} + (\overline{6} \cdot \overline{1})$ is not provable anymore, something we have already seen in the deduction of multiplication* above. This lack of provability in turn means that the resulting model of arithmetic is

[2] To make the distinction between numbers and numerals explicit, from now on an overline will be added for the numerals. For example, $\overline{0}$ is the name of the number 0.

not isomorphic to the standard interpretation of arithmetic, i.e the arithmetic about the natural numbers, anymore (Boolos et al., 2007, p. 283). We will set aside the implications of this for now, and address those in Sect. 8.4.

An alternative approach is to keep the otherwise intended interaction between multiplication, addition, and succession the same. To this end, it can be observed that multiplication is defined in terms of addition. So, when multiplying, one applies the function of addition as well. If the derivation of the above mentioned contradiction is to be avoided, then changing multiplication implies also changing addition, otherwise $\overline{6}+(\overline{6}\cdot^{*}\overline{1})$ would equal $\overline{12}$, and not $\overline{13}$. Thus, in this approach, the newly suggested function multiplication* requires a different function of addition, say addition* (or $+^{*}$), to deal with the stipulated changes to the number 13 in the mathematical antecedent of the counterfactual. Working through the example, we get $\overline{6}\cdot^{*} S(\overline{1}) = \overline{6} +^{*} (\overline{6}\cdot^{*}\overline{1})$, and therefore $\overline{6}\cdot^{*}\overline{2} = \overline{6} +^{*} \overline{6}$, and $\overline{13} = \overline{13}$.

Of course, the changes to addition* also need to be accommodated in the recursive definition for addition itself (axiom 5): $\overline{6} +^{*} S(\overline{5}) = S(\overline{6} +^{*} \overline{5})$. But this leads to $\overline{6} +^{*} \overline{6} = S(\overline{11})$, and as a result $\overline{13} = \overline{12}$. To prevent the contradiction popping up again here, a change in the successor function is then also needed. Let us call it 'succession*' or S^{*}). We get: $\overline{6} +^{*} S^{*}(\overline{5}) = S^{*}(\overline{6} +^{*} \overline{5})$, with which we get $\overline{6} +^{*} \overline{6} = S^{*}(\overline{12})$, and in the end we get the desired equality $\overline{13} = \overline{13}$.

The recursive character of the functions of multiplication (in terms of addition and succession) and addition (in terms of succession) thus results in 'downward' changes on this approach. And we see that both functions effectively supervene on the successor function. Taken together, and again ignoring the base case as we did before, this means that the recursive step in the function of multiplication will also have to accommodate these changes:

$$\forall x \forall y (x \cdot^{*} S*(y) = x + *(x \cdot^{*} y))$$

In contrast to the first approach, this way of addressing the contradictions thus requires some changes to be made to some aspects of number theory other than multiplication.

Depending on how to decide which contradictions are relevant and which are not, the hope is that one should be able to draw a line somewhere and allay the worry of the local twiddle reverberating through the whole of number theory. An additional aspect is that we have to be sure that we speak about the very same thing in the counterfactual situation as in the factual situation, as the counterfactual has to be about this something or other. Thus in the case of Suzy, we have to make sure that we are speaking about the selfsame person in both the factual and counterfactual situation. And in the case of the number 13 in multiplication* (that is, with 13 being not prime), we have to make sure we are speaking about the very same number as the number 13 in the function of multiplication. For example, it has to be so that the number 13 in multiplication* still has all the (relevant) other properties of the number 13 in multiplication, for instance of being between 11 and 13. Baron et al. (2017, p. 5), speaking about the number 17 in what follows, state that:

> in exploring what would happen if 17 were composite, there is no possibility that we're talking about something other than 17 itself, because it is simply part of the specification of the case to be considered that we are talking about 17.

They compare this to a remark by Kripke (1980, p. 44) in which he says that "Possible worlds are stipulated, not discovered by powerful telescopes. There is no reason why we cannot stipulate that, in talking about what would have happened to Nixon in a certain counterfactual situation, we are talking about what would have happened to him". But it has to be noted here that Nixon could have existed with very different properties, whereas it is not clear at all if this could also be the case for, say, the number 13. It might seem quite trivial in the case of Nixon, but number theory is somewhat particular in this respect. Indeed, the intention is to specifically speak about the number 13 in the counterfactual situation. But when ironing out the (relevant) contradictions, we have to make sure that we do not do it in such a way that we end up talking about something else than what we intended to talk about in the first place. The question as to what it is to be a natural number, and whether we indeed speak about them in the counterfactual situation, will be discussed in the next chapter.

7.3 Conclusion

Starting with a concrete suggestion in the literature on how to capture the explanatory role of mathematics counterfactually in Sect. 7.1, it was noted that two issues arise that suggest a possible shift in context which results in a structure different than that of the natural numbers. These were discussed in Sect. 7.2. First, by changing the function of addition in what we called an 'exceptionalist' way, in which the newly begotten model of arithmetic will not be isomorphic to the standard interpretation anymore. The relation of the standard model of arithmetic to the structure of the natural numbers will be further discussed in Chap. 8. Second, by accepting the downward changes when changing the function of multiplication, it was shown that this ultimately requires a change in the successor function S. The role of the successor function is however of special importance in determining the context in which the natural numbers are to be understood. In light of Peano axiom 3, the induction axiom, it is clear that it is this particular axiom that makes some property P to hold for all elements of the set (of natural numbers). In the case of number theory the essential property P of the set is the successor function S. In other words, it appears that to be a natural number is to be a member of the set containing the successor function. Therefore, changing an apparent local thing in the multiplication function turns out to have the undesirable consequence that the natural numbers are not closed under succession anymore. The further implications of this observation will also be discussed in Chap. 8.

Chapter 8
De Re Knowledge of Numbers

Abstract Based on insights from discussions on computation, the structure of natural numbers, de re knowledge of numbers and the standard interpretation of arithmetic, it is questioned whether a counterfactual account of explanation can work for number-theoretic counterpossibles. Making use of Peano arithmetic, emphasis is put on the role of the standard induction scheme, as it contains the symbol for the successor function, as such suggesting an intimate connection between Peano numerals and de re knowledge of numbers. Because of the resulting shift in context in the antecedent, the way a contexts stops an explosion of ramifications in the case of ordinary counterfactuals might not be available in the case of number-theoretic counterpossibles.

With the need to accommodate the changes of the new function multiplication*, either by means of an 'exceptionalist' definition, or by having to change the successor function *S* as a result of the acceptance of downward changes from the initial changes to multiplication, we enter murky waters. The worry is that in both cases the changes that are intended to be invoked locally, or at least sufficiently local, lead to a change in the structure of the numbers, something which will ultimately result in a structure that is different than that of the natural numbers. And if this were indeed the case, then doubt is cast on the requirement that the mathematical content of the counterfactual should be same, in the relevant aspects, to the mathematical content in the factual situation.

To introduce the what is at stake, Sect. 8.1 discusses computation of number-theoretic functions and its relation to the symbols used in those calculations. These symbols, the numerals, appear to have a special status compared to other names for numbers. That is, they are part of a number system as a whole, and as such allow for existential quantification in intensional contexts. Quantification in intensional contexts is the topic of Sect. 8.2. Section 8.3 will build on the previous section, and point to the importance of the idea of succession, focussing on the Peano numerals, and its link with de re knowledge about the natural numbers. Section 8.4 then moves on to the discussion on the intended models of arithmetic, where the similar observation is made that numbers are given collectively. They are said to instantiate

© The Author(s), under exclusive license to Springer Nature Switzerland AG 2025
L. A. Tump, *On Mathematical Explanations of Empirical Phenomena*,
Synthese Library 517, https://doi.org/10.1007/978-3-031-98860-8_8

a unique structure, and a change in the structure of the natural number system results in a system that is not isomorphic to the standard model anymore.

Thus far, the focus will have been on the antecedent of the counterpossible. Section 8.5 shifts the focus to the consequent, and argues that due to the shift in structure, the meaning of the empirical begotten consequent will differ from the antecedent if certain changes to the natural numbers are made. Also, as mentioned in Sect. 6.2, the counterpossible has to be non-trivially true for it to be able to play a part in an explanation at all. Resorting to a contradiction tolerant logic might solve the issue of the trivial truth of all counterpossibles, but in such a system the idea of succession is let go. This is something which, in light of the earlier chapter, can be considered to be a sufficiently severe problem in itself. Finally, Sect. 8.6 summarizes the chapter and concludes that the changes made in the counterpossible are problematic for an account of explanation, if the account is to be part of an argument for realism about natural numbers.

8.1 Computation, Numbers, and Numerals

To get a better grasp on what is at stake when we speak about the structure, or sometimes called context, of the natural numbers, let us first turn to the domain of computation in order to highlight an important observation about the relation between numbers and numerals. This will provide us with some insight about what we briefly touched upon in Sect. 7.2, namely that the Peano axioms contain the numeral for zero and for the successor function. These numerals *denote* the (Peano) numbers, and given the fact that the numerals denote the (Peano) numbers, it should, at some point at least, be clear what the referent of a numeral is.

The concept of computability is mostly thought of as an intuitive notion. It primarily pertains to functions, and the functions themselves relate members of one set uniquely to members of another set. A function is said to be computable if there exists, in principle, a finite (mechanical) procedure which computes the output values of the function given certain input values (Rescorla, 2015, p. 338). This may sound quite clear, but as Boolos et al. (2007, p. 23) succinctly put it with regard to the Church-Turing thesis, which roughly states that every effectively computable function on the natural numbers is Turing computable, that is, it is computable by a so-called 'Turing machine':

> "[t]his thesis is not obvious, nor can it be rigorously proved (since the notion of effective computability is an intuitive and not rigorously defined one), but an enormous amount of evidence has been accumulated for it".

The Turing machine is indeed a well-known formalization for computable functions, and it is much used for the study of number-theoretic functions. It is a theoretical machine, in the sense that there are no practical limitations regarding time or storage capacity and the like. Effectively, Turing machines carry out computations, and they do so by following a set of explicitly defined rules which are jointly called a 'procedure'. The exact workings of such machines, although interesting in itself, are not relevant here, but what is important to notice for our current discussion, is that these computations are only carried out on the positive integers. The positive integers, in turn, are represented by some notation, i.e. symbols (Boolos et al., 2007, p. 25). As such, numbers themselves are not encountered in the computations at all. Rather, the computing of number-theoretic functions is done via, or mediated by, the notational systems employed in the algorithms (Shapiro, 1982, p. 14). Thus, the arguments in the computation are given by means of the numerals of some notational system or other, e.g. unary, binary, decimal, etc. Therefore, the initial observation is that we seem to compute number-theoretic functions via the numerals, rather than directly over the numbers. This brings up the question what the relation between numbers and numerals is.

At this point, it might be objected that the computation of number-theoretic functions mediated by notation is still computation over number-theoretic functions. Because in the end it could be said that "[m]ediated computation is still computation" (Rescorla, 2007, p. 273). Additionally, we are probably mostly interested in number-theoretic calculations anyway, and not so much in the notational systems of those calculations (Rescorla, 2007, p. 253; Shapiro, 2017, p. 269). This, however, rather restates the need for an account of the relation between the syntactic and the numerical. And of course, in that sense we are also interested in the notational systems. After all, we want to know which number a certain notational representation is about. It is therefore that Kripke (2011, p. 344) states that

> the notion of computability is best seen as one of having a procedure for knowing which *number* is the value of the function [my emphasis].

This implies that at some point, after doing some calculation, the sense of a term should be immediately revelatory.[1] Whatever is begotten at that point is something that stops the buck, as Kripke (2011, p. 261) calls it. In other words, the outcome of a calculation is a 'buckstopper' when it is the case that, at that point, no further calculations are required to get to the referent of the term. The question is then which numerical terms can be considered to be buckstoppers? Kripke gives the following example (p. 261). Given some computable function $f(n)$, say 'the square of n', then 'the square of $\overline{3}$' is not immediately revelatory, for some calculation can still be carried out to get to the referent of that description. Or, to put it differently, it still makes sense to ask 'what number is 'the square of $\overline{3}$'? But $\overline{9}$, on the other hand, is a buckstopper (for people used to the decimal system in any case). No further calculations are needed for anyone familiar with (the decimal) numbers. To emphasize the importance of the use of numerals in computations (and in the Peano axioms), the observant reader notices that in this example the numerals for

[1] Given Kripke's interest in the work of Ludwig Wittgenstein, it is worth mentioning that the idea that only when the outcome of a calculation is immediately revelatory it is a buckstopper, is similar to Wittgenstein's requirement that a proof, i.e. the calculation and the outcome, must be perspicuous (Wittgenstein, 1956, p. 65e).

the numbers are used (i.e. $\overline{3}$ and $\overline{9}$). Compare the similar remark by Boolos et al. (2007, p. 24):

> if the value of the function we are trying to compute is a number, what our computations in fact end with is a numeral for that number.

Given that the numerals denote the numbers, the context in which the question 'what number is this?' is asked matters. Some notational system might be revelatory to some, but not to others. This brings us to the more general problem of de re propositional attitudes, specifically with regard to what names for numbers are eligible for existential quantification in intensional contexts, and which notational system is (or which notational systems are) acceptable.

8.2 Numbers and De Re Propositional Attitudes

Ackerman (1978) aims at determining the conditions under which it can be established what arithmetical terms are eligible for existential quantification in intensional contexts, specifically the integers.[2] This matches well with the present investigation, since we are primarily interested in the structure and status of the natural numbers.

Consider the following example from Quine (1956, p. 178) with two sentences that express a de re and a de dicto belief respectively:

(1) $\exists x$ (Ralph believes x is a spy)
(2) Ralph believes that $\exists x$ (x is a spy)

Sentence (1) states that there is someone (i.e. a specific person) whom Ralph believes to be a spy, whereas sentence (2) states that Ralph believes that there are spies (in general).

Kaplan (1968, p. 192) then asks us to suppose that Ralph believes that no two spies are of the exact same height, and that the shortest spy is a spy. The description 'the shortest spy' takes the place of x in 'x is a spy'. The question is now whether, from Ralph's supposed belief about the height of spies, we can infer (1). In other words, whether existential generalization is justified in relation to the description 'the shortest spy'. The suggestion is that the description with which x was substituted is not suitable for such existential generalization. After all, it is possible that Ralph, like most of us, is not familiar with any particular person who is a spy.

Now substitute the name 'Ortcutt' for x, so that Ralph believes that Ortcutt is a spy. Kaplan (1968, p. 192) suggests that in this instance (1) can be inferred, under the condition that the name 'Ortcutt' singles out the person Ortcutt only. In other words,

---

[2] Ackerman specifically targets the integers, because she does not have firm intuitions about other kinds of numbers (Ackerman, 1978, p. 145).

the relation between the name and the person has to be of a specific kind, such that the name represents the individual uniquely. Kaplan calls such names 'vivid' (Kaplan, 1968, p. 201). It must be mentioned here that Quine thought that in the end no principled distinction could be made due to the context sensitivity of the terms and descriptions, at least in the examples he discussed and which are similar to the ones mentioned here (Rescorla, 2015, p. 359). It is also important to realize that these examples do not (yet) include an exegesis with regard to arithmetical terms, although Kaplan (1968) does go on to speak about arithmetical terms.

Moving the discussion on toward arithmetical terms, Ackerman (1978) subsequently asks whether the same would be the case if 'x is a spy' were substituted for, for example, 'x is a perfect number' (a perfect number is a number equal to the sum of its divisors). Thus, the question is whether all arithmetical terms are eligible for existential quantification in doxastic contexts. Similar to the earlier two sentences, we get the following:

(3) $\exists x$ (Ralph believes x is a perfect number)
(4) Ralph believes that $\exists x$ (x is a perfect number)

She now applies the same procedure as Kaplan (1968), and substitutes 'the smallest perfect number' for x in 'x is a perfect number'. The first situation supposes that Ralph has heard of perfect numbers, and that he believes that the smallest perfect number is indeed a perfect number. Ackerman concludes that (3) cannot be inferred from this supposed belief: it is after all possible that Ralph does not know of any specific number that it is perfect. Contrast the just mentioned belief with a second situation in which Ralph believes that 6 is the smallest perfect number (with 6 substituted for x in 'x is a perfect number'). In this case, she argues, the inference to (3) is justified because it is a de re belief about a number (Ackerman, 1978, p. 147). The point is that in the first situation, the argument would have carried over content from the de dicto sentence to the de re sentence, were the inference accepted. According to the counterexample, however, this should not be the case. The situation is indeed different in the second case.

This leads her to the observation that there is something special about what is expressed by numerals as opposed to other names (for numbers). The suggestion is that a numeral

> *directly* specifies the position of its referent in the progression of numbers (Ackerman, 1978, p. 151).

The difference with a term such as 'the smallest perfect number', which in some sense also specifies the position of its referent in the progression of numbers, is that knowing to which position the numeral itself refers to is simply a matter of understanding the numeral. In the case of a term such as 'the smallest perfect number' some more mathematical knowledge is needed than or what is needed in order to understand a certain numeral. For one, it has to be understood what a perfect number is.

Concerning the specification of the position of the referent of a numeral in the progression of numbers, we remember the observation made at the end of Chap. 7

regarding the fundamental role the successor function appears to play. What is of importance here, as was observed by Ackerman, is the relation between succession and position, in casu the next position. A progression can typically be thought of as an order with an initial element, such as can be seen in the natural numbers: $0, 1, 2, \ldots$.

Given these observations, the question is thus whether it makes sense to speak about changing something to a number in isolation, such as the twiddles to the number 13, if that would mean a change to the successor function. So far, the tentative suggestion is that this is problematic, and that numbers are indeed to be understood holistically. This is why Ackerman observes the following, when she says that the referent of a numeral can be known directly by understanding the numeral itself:

> [numbers] can be understood only in the context of a system of numbers, and knowing and understanding a system of numerals seems to be a matter of knowing how to generate in order the progression of numerals and knowing how to count transitively ... in accord with the progression (Ackerman, 1978, p. 151).

With transitive counting is meant counting the cardinality of a collection, in such a way that the numerals for the natural numbers are generated in order, and each numeral, of whichever notational system, is associated with a unique object (Shapiro, 2017, p. 272). The next section will discuss the importance of the successor function, and the relation it has to progression and position.

8.3 The Importance of the Successor Function

When looking at the idea of succession, let us keep the Peano induction scheme (3), as mentioned in Sect. 7.2, in mind, and remind ourselves that it contains the symbol for zero, 0, and the symbol for the successor function, S.

When discussing de re propositional attitudes about natural numbers, Shapiro (2017) stresses, like Ackerman (1978), the importance of the successor function in our understanding of natural numbers, and highlights the importance of the successor function on the numbers themselves (Shapiro, 2017, p. 279). Take for example unary notation, where, for instance, the numerals are the sequences of strokes such as used in a simple tallying procedure. The number n would then be denoted by n strokes (e.g. the number seven would be |||||||). This is at the same time a structured order: one stroke after the other, say, from left to right. This means that, relative to this notational system, the successor function is computable. This is indeed an important aspect of the natural numbers when they are used for, for example, determining the cardinality of some set in answer to the question 'how many?'. In general, the cardinality is begotten by reciting, or rather enumerating, the numerals of some notational system (Rescorla, 2007, p. 269).

In fact, Shapiro (2017, p. 279) goes on to highlight two features of unary notation, such as can be seen in the previously marking down of strokes, of which he thinks that it makes unary notation privileged over other notations. The first is that each numeral for some number n *is* a collection (of strokes) with cardinality n. The intimate relation between the numeral and its corresponding cardinality can be seen directly. The second feature of unary notation is that it displays the order of the collection (as mentioned above: one stroke after the other, say, from left to right). Thus, unary notation is privileged because it "displays both the cardinality and the corresponding ordinal" (Shapiro, 2017, p. 279).

This does of course not mean that all numerals in unary notation are buckstoppers, at least not for ordinary humans. The unary numeral for a very large number would most likely still prompt the question 'what number is this?'. But this is not only a problem of unary notation. At some point, no matter what notational system is used, there will be some numerals that are not buckstoppers. The numerals simply get too big (Steiner, 2011, p. 165). However, when computability theory is concerned, one idealizes away from the limitations of ordinary human beings (capacity, time, etc.). From the computational point of view, the computer is an idealized human being (Shapiro, 2017, p. 277). Given the lack of restrictions of this kind for such idealized agents, very large numerals can therefore still be buckstoppers.

Returning to the issue of the importance of the successor function, and an initial element on which this function operates, it is clear that the Peano numerals display both elements directly. In a sense this should come as no surprise, because the Peano axioms were meant to define (or describe) the natural numbers. They capture the idea of succession in a fundamental sense, and it uses numerals, i.e. the Peano numerals, to represent the natural numbers. In fact, the successor function is hard coded, so to speak, in the Peano numerals themselves. This can easily be seen, because the Peano numerals for the sequence of natural numbers are $\overline{0}, S(\overline{0}), S(S(\overline{0})), \ldots$. For every $n > 0$, the number of S's in the Peano numeral $\overline{n}$ has cardinality n. In light of Shapiro's discussion of unary notation, it is seems that putting strokes on paper in an orderly fashion, as we have seen above, already gives us something similar to the first two Peano axioms. The first stroke represents the number one, two strokes represent the number two, and so on. The number zero is represented by the empty string, i.e. nothing is jotted down (Shapiro, 1982, p. 14).

This makes the Peano axioms particularly relevant in the current context, with specific emphasis on the standard induction scheme, axiom (3), which contains the symbol for the successor function S, and the symbol for 0. There is evidence that this scheme plays a key role in the status of the Peano numerals (Heylen, 2014, pp. 20–21): given the standard induction scheme (with added knowledge operator) one can deduce that the Peano numerals allow for existential quantification in epistemic contexts.[3] In other words, the Peano numerals are canonical. Or, in Kripke's words again, they are 'buckstoppers'. And importantly, without the standard induction

[3] Relating this to the topic of Part I, the explanation, or potential explanation, is a deductive explanation where the claim that the Peano numerals allow for existential quantification in epistemic contexts is deduced from the full induction scheme.

scheme one cannot deduce the canonicity of the Peano numerals. Take for instance the earlier mentioned difference between 'the square of $\overline{3}$' and $\overline{9}$. Analogous to the epistemic puzzles in Sect. 8.2, if someone knows that 'the square of $\overline{3}$' equals 'the square of $\overline{3}$', this does not entail that they know that 'the square of $\overline{3}$' equals $\overline{9}$. A similar observation, but then in relation to algebraic investigations, about the importance of a canonical form is also discussed by Lawrence (2021). The idea there is that, when determining the value of a variable in an equation, an answer needs to be a complete answer. This means that, in the context of giving solutions to algebraic equations, the question as to what value a variable has would not be appropriate given that the context speaks of a solution to an equation.

The canonicity of the Peano numerals suggests that they are intimately related to de re knowledge about natural numbers. The point is that once the induction principle is accepted, the canonicity of the terms follows. Similarly, if instead of the Peano numerals the induction was formulated by, say, $+\overline{1}$, then the canonicity of the following terms would follow: $\overline{0}$, $(\overline{0}) + \overline{1}$, $((\overline{0}) + \overline{1}) + \overline{1}$ (Heylen, 2014, p. 12). If we recall Ackerman's remark that understanding numbers (and a system of numerals) is very closely tied to knowing how to generate a progression, then it appears to be the case that some form of induction is already implicit in our very understanding of (the natural) numbers: you know the numbers if you are able to get to the next number, starting from zero. This so-called 'intransitive counting' requires an understanding of a recursive procedure for generating the progression (Benacerraf, 1965, p. 50). Understanding a number, then, is to understand its place in a progression. And if this observation is right, then a number cannot be separated from the system of numbers of which it is part.[4]

Relating these observations to the counterpossible introduced in Chap. 7, the importance of them becomes clear. Given the counterpossible, the intention is to speak about the natural numbers themselves, thus de re. It is not the intention to speak about their descriptions, i.e. the numerals. And because different terms for numbers do not necessarily all lead to de re knowledge of numbers, the primary interest is in the canonical terms for the numbers. If one were intent on separating the individual numbers from the system of numbers, then it is not clear how this can be done without losing a basic tenet of our understanding of the natural numbers. In reference to Priest (2016, p. 2660), when we imagine that 13 is not prime (and it is), we intend to imagine something about that very number.[5] However, change the number system, and you get different numbers with different canonical properties, etc. And if this is correct, then stating something like $\overline{2} \cdot^{*} \overline{6} = \overline{13}$ results in the fact that the set of elements does not equal $\mathbb{N}$ anymore.

[4] When buckstoppers are concerned, if indeed numbers cannot be separated of the system it part of, in some sense it seems odd to draw a line somewhere between numerals that are considered to be buckstopper and numerals that are not considered to be buckstoppers.

[5] His original formulation is: "And when I imagine that 361 is a prime number (it isn't) I am imagining something about that very number." (Priest, 2016, p. 2660).

8.4 The Intended Models of Arithmetic

Halbach and Horsten (2005, p. 181) also endorse the view that numbers are given collectively, but approach it from the perspective of the standard interpretation of arithmetic, and specifically the functions of addition and multiplication. To be more precise, the issue they are concerned with is not solely about how and why a numeral $\overline{n}$ for some number n indeed refers to the number n. Rather, they are concerned with the question how a single structure for doing arithmetical operations on the natural numbers can be singled out (Horsten, 2012, p. 276).

As the problem was stated in Sect. 7.2, if the function of multiplication* is redefined in what was called an 'exceptionalist' way, by for example taking multiplication for all m, n other than $\overline{2}$ and $\overline{6}$, and then define $\overline{2}$ and $\overline{6}$ separately, such that $\overline{2} \cdot^* \overline{6} = \overline{13} = \overline{6} \cdot^* \overline{2}$, then it is the case that the resulting model of arithmetic is not isomorphic to the standard interpretation of arithmetic anymore. This is relevant, because the standard model of arithmetic is the model of arithmetic intended to be about the natural numbers.[6] The objects in the intended models are (the referents of the names for) the natural numbers (Boolos et al., 2007, p. 104). The point is that the objects in the standard model constitute not just any structure, but a specific one, namely the structure for doing arithmetical operations on the natural numbers. And in case there are two or more models of it, then there is an isomorphism between these models instantiating the same structure. First-order Peano Arithmetic, as is well known, has of course non-standard models, i.e. models that are not isomorphic to the standard model, and in some of which the Peano axioms are satisfied nonetheless. These models are not intended and need to be ruled out.

To rule out the non-standard models, Benacerraf (1965, pp. 52–53) expressed the requirement, amongst others, that given two objects in a progression, there should be a decision procedure for determining which of those is the greater of the two, and that in order to be able to do so, the relation 'greater than' needs to be recursive.[7]

However, the recursive character of the 'greater than' relation is not sufficient to rule out non-standard models (Halbach & Horsten, 2005, p. 178). There is at least one non-standard model with a recursive order relation, i.e. the 'greater than' relation.[8] It is therefore suggested that (either of) the operations of addition and multiplication should also be recursive. One reason for this is the fact that being

[6] Some might prefer to refer to it as 'standard models', for instance a structuralist, since there can be various models of arithmetic instantiating the same structure. The philosophical question here, and elsewhere, is whether isomorphism implies identity.

[7] The other two requirements were the identification of some sequence of order type ω as the numbers, and specify the cardinality relation such that a set of some cardinality is correctly related to its cardinal number, i.e. to an element of the sequence (Benacerraf, 1996, p. 189). Halbach and Horsten (2005, p. 175) are quick to point out that the requirement of an ω-sequence begs the question, because the notion of an ω-sequence is usually defined as a sequence which is isomorphic to the Von Neumann ordinals (for the Von Neumann ordinals, see Sect. 3.2.1).

[8] See Boolos et al. (2007, p. 305, or section 25.1), for a discussion on the topic.

able to calculate the sum and product of two objects seems to be fundamental for these objects to be considered as numbers. Besides, adding the requirement of recursion for the functions of addition and multiplication does indeed rule out non-standard models. This is due to Tennenbaum's theorem, which states that if M is a countable model of first-order Peano arithmetic, and the operation of addition (and/or multiplication) is recursive, then the model M is isomorphic to the standard model (Button & Smith, 2012, p. 114).

Button and Smith (2012, pp. 116–117) observe that an argument for singling out the standard model that makes use of Tennenbaum's theorem can be challenged. The problem is that the theorem makes use of the notion of 'recursive function', i.e. a function that is computable in some finite number of steps. To understand what a recursive function is, it needs to be understood what a finite number of steps is, for instance for writing down strokes in unary notation. This would, it is argued, presuppose knowledge of numbers, or to be more precise, knowledge of a finite number. A similar observation is made by Halbach and Horsten (2005, p. 180), who distinguish between a theoretical and a practical notion of computability. The former is a notion of pure mathematics and applies to functions and relations of numbers. Therefore, the theoretical notion cannot be used, precisely because it presupposes numbers. A practical notion of computability would not rely on the prior use of numbers, because it is about manipulating the symbols employed in the algorithms (cf. Shapiro's observation mentioned in Sect. 8.1, that computations are done via the numerals.). In other words, when we follow arithmetical operations, we work through the symbols. The same goes for instance for Turing machines, when actual computation is concerned (Halbach & Horsten, 2005, p. 181).

Because there are different notational systems possible, the idea is not to identify a single notational system for the natural numbers. Rather, any notational system that is isomorphic to the standard model of arithmetic will qualify, as long as the notational system satisfies the Peano axioms. This way different notational systems can be used without favouring one or the other (e.g. unary, binary, decimal, or perhaps other numeral systems such as Roman numerals). For the structuralist account of Halbach and Horsten this is important, because identifying a single, or canonical, notation requires the ability to distinguish between a standard and non-standard notation. And this effectively restates the problem of distinguishing standard from non-standard models, because *that* problem now depends on the notation being standard or non-standard (Halbach & Horsten, 2005, p. 176). Relying on the notion of practical computability is meant to side step this issue. To be able to distinguish between the different symbols in computations is a necessary requirement for there to be computations. Without this ability, algorithms could not be used, for the instructions would be ambiguous. The combination of the possibility of performing the standard arithmetical operations on a notational system, together with Tennenbaum's theorem leads to the following conclusion by Halbach and Horsten (2005, p. 183):

> [the] intended models [of arithmetic] are notation systems with recursive operations on them satisfying the Peano axioms.

This is, of course, not an uncontroversial thesis. The step from the fact that we manipulate symbols when doing arithmetic to the statement that the intended models *are* notation systems with recursive operations on them is not straightforward. However, for the present purposes, this thesis need not be defended here. What is of importance, is that there appears to be a strong connection between de re knowledge of numbers and computation.

But the criticism does not stop here. Button and Smith (2012) observe that with regard to the practical notion of computability, it is just as hard to distinguish between standard and non-standard arithmetical operations as it is to distinguish between standard and non-standard numerals. Given that the practical notion of computability is concerned with our ability to manipulate notational systems, and we (or computers for that matter) are limited in our abilities, questions about what could be computed in principle would go out of the window. To speak about 'in principle' is again to speak about the earlier mentioned arbitrary finite number of steps. Given this, operations on tractably small numbers are unproblematic, but in principle operations are not (Button & Smith, 2012, p. 118). It must be noted that for the context of the problem currently discussed, i.e. counterpossibles about numbers in relation to empirical facts such as the number of years of the life cycles of Periodical Cicadas, we are concerned with tractably small numbers.

Returning to the newly suggested function multiplication*, with m, n other than $\overline{2}$ and $\overline{6}$ the same as multiplication, and $\overline{2}$ and $\overline{6}$ defined separately, it is clear that the resulting model is not a model of Peano arithmetic, because $\overline{6} \cdot^* S(\overline{1}) = \overline{6} + (\overline{6} \cdot^* \overline{1})$ is not provable anymore (cf. Sect. 7.2). This means that there is at least one non-standard model, and that therefore the resulting model is not isomorphic to the standard interpretation. From this it follows that, since the standard models are the intended models about the natural numbers, the new model is not about the natural numbers.

One more remark with regard to the above mentioned requirement of recursion, because the 'exceptionalist' definition of multiplication* is also recursively defined. But recursion is in itself not sufficient. The notational system must also be a structured order (cf. the remark of Shapiro about unary notation in the previous section) in which the arithmetical operations, or at least simple functions such as addition and multiplication, are computable. Horsten (2012, p. 279) gives the following example with regard to the Hindu-Arabic numerals (which is a recursive notational system). If permutations are made to the numerals in such a way that the standard succession function breaks down (e.g. $\overline{12}$ is now the successor of $\overline{231}$), then the simple arithmetical operations of addition and multiplication are not computable anymore. It must be noted that although the suggested function multiplication* is both recursive and computable, the resulting model does nonetheless not instantiate the standard model. This is the reason why it is required that the Peano axioms must also be satisfied.

The counterpossible about the number 13 not being prime is not the only example of a counterpossible regarding the natural numbers given by Baron et al. (2020). In another, but similar, example of a twiddle to the natural numbers, the same problem occurs. This time, it is about numbers counterpossibly being even or uneven, while they are not. Here too, it is very clear that the structure of the numbers is actually changed, and moves away from that of the natural numbers. Consider the following counterpossible:

> ... twiddle the natural numbers by making it so that none of 503, 504, or 505 is even. The twiddle ramifies as follows: the product of two natural numbers is even (if and) only if one of the numbers is (Baron et al., 2020, p. 29).

What is mentioned here is supposed to be about the *natural* numbers, just as it was with the counterpossible about 13 not being prime. But given the earlier observations, if 504 is considered to be not even, this means that the standard arithmetical operation of dividing it by 2 can not be performed on it. One of the ramifications is that the underlying structure does not satisfy the Peano axioms anymore. In other words, reiterating the argument from earlier, the model that results is not isomorphic to the standard model of arithmetic. And since the standard model of arithmetic was supposed to be, or intended to be, about the natural numbers, again, what results is a structure different from that of the natural numbers. More so, given the standard induction scheme, Peano axiom (3), and the functions for addition (4 and 5) and multiplication (6 and 7), it follows that of three consecutive numbers, at least one needs to be even (or uneven, depending on the initial number of the series of three). In the counterfactual situation this is not the case, by stipulation, and thus we see that in this case too, the structure is changed and subsequently differs from the structure of the natural numbers.

These observations suggest that changing an apparently local thing in a system of numbers, such as 13 factorizing in 2 and 6 as well as in 1 and 13, or making none of 503, 504, 505 even, has the consequence that one ends up talking about two different structures. One of which is the natural numbers, the other some other kind of numbers. And this very fact would in turn mean that the supposed counterfactual is actually not a counterfactual containing a *natural* number, in our main example being the prime number 13 such as used in the original function of multiplication. This is so, because the numeral 13 now *means* something different than the numeral in the function multiplication*, resulting from the changes to the structure.[9] In an argument for realism about natural numbers, this is very problematic indeed.

---

[9] A similar idea can already be found in section 6 of Quine (1986), titled 'Change of logic, change of subject'. Now, someone like Williamson (2007) might argue that the focus on succession as a requirement for the subject matter of mathematical counterfactuals is misplaced, at least in an *epistemic context*, and that one could still meaningfully speak about the natural number 13 in the counterfactual situation, analogously to his discussion on modus ponens. Although an inferentialist might argue that modus ponens is constitutive of the meaning of the conditional. Disagreement between experts about such issues can of course occur, but it leaves open the question whether this can occur in the same sense in the specific case of number theory, because of the close connection between the structure of a number system, or the idea of succession, and numbers.

8.5 Further Ramifications of the Local Twiddle

The focus so far has been on the antecedent of the counterfactual, but the consequent is also of importance. If the meaning of the antecedent changes, how is it then supposed to latch on to the empirical content of the consequent? In a sense, with multiplication*, we lost the tallying procedure for the measuring of the actual life cycles. Without it, how are we to decide on the cardinality of the set of these life cycles? To remind ourselves, a tallying procedure, even as simple as putting strokes on a piece of paper, already gives us something that is isomorphic to the Peano numerals. And it does so, because it gives you an initial element and a procedure of succession.

The worry about the latching on of the antecedent and consequent, due to a difference in context between both sides of the counterfactual, is a similar kind of worry to the one raised in Sect. 4.1.2 in relation to the EIA, from which the case of the mathematical explanation of the prime life cycles of the Periodical Cicadas is taken. Pertaining to the counterpossible at hand, i.e. that if 13 were not prime, then Cicadas would not have had life cycles with a 13-year length, two issues need to be addressed. The first issue pertains to the difference in meaning between the antecedent and consequent after the change to the antecedent. The second issue pertains to the fact that for a counterpossible to be able to be used in an explanation, it should be non-trivially true (cf. Sect. 7.1). These two issues come together in the problem that will be formulated in this section.

As already mentioned in Sect. 2.3, the EIA is in essence an inference to the best explanation. And to get it off the ground, the explanans as well as the explanandum need to be assumed true. For this reason, the reference to numbers in the explanandum, due to the notion of primality in it, needed to be paraphrased away in first-order logic with identity to prevent circularity (Bangu, 2008; Baker, 2009). It was then observed that the context between the explanans and the explanandum differs after this modification. The explanans treats of an entity, i.e. the number 13 (or 17), whereas the explanandum treats of a plurality, i.e. the collection of objects of the sentence in first order logic with identity. Hence the observation that a bridge principle is needed. It was then suggested that HP, given the strict but non-trivial requirement that it must not be an analytical principle, could link the entities to the pluralities.

Similarly, to bridge the difference in meaning in the case of the counterfactual statement about the number 13, thus to hook the modified antecedent to the empirical content of the consequent, some sort of criterion is needed too. This criterion needs to ensure that there is at least some overlap between the two meanings on either side of the counterpossible. Without overlap in the relevant sense, specifically for conditionals that are not supposed to be trivially true, there is no connection between the two sides. But this overlap is needed to ensure that the counterfactual can be used in the explanation of the explanandum in the first place.

Let us start with the second issue, pertaining to the required non-trivial truth of the counterpossible for it to be usable in an explanation. Following tradition,

Lewis's closeness-based semantics might be suggested as a criterion for relevance for the evaluation of counterfactuals (Baron et al., 2020, p. 4). As already discussed in Sect. 6.2, on this account closeness is cashed out in terms of similarity. Lewis also speaks about intrinsic properties, which can be used to compare possible worlds. He defines intrinsic properties as follows:

> Property P is intrinsic iff, for any two duplicate things, either both have P or neither does (Lewis, 1983, pp. 355–356).

Relating this definition of intrinsic properties to the notion of similarity between worlds, then different worlds are (intrinsically) similar when they share some intrinsic property P.

One of the problems with the standard semantics of possible worlds is that any world with inconsistencies, for instance any world in which standard mathematics is false, is an impossible world. This is also the case for worlds in which the Peano axioms are violated. After all, the suggested counterpossible is in contradiction with the Peano axioms. As a result, since the antecedent harbours the contradiction it does, the counterpossible comes out as trivially true (cf. Chap. 6). As of course would any other counterpossible. If this triviality is to be avoided, inconsistent worlds need to be accommodated in the semantics of counterfactuals. In order to do so, it could be suggested that Lewis's account of closeness to be extended to encompass, besides possible worlds, impossible worlds too (Baron et al., 2020, p. 4).

Thus, in the case that the changes to the number 13 lead to contradictions given the standard axioms, and these contradictions cannot be contained as hoped, then it could indeed be accepted that a contradiction tolerant logic be used (Baron et al., 2020, p. 5). Such a move would allow for some relevant aspect in the relation between the antecedent and the consequent to be maintained in case of inconsistent worlds, without making the counterpossibles trivially true. In this situation, it would not be the case that everything is true at such an inconsistent world, including the negation of everything.

However, some observations about resorting to a paraconsistent logic in relation to the present context need to be made here. First, it is certainly true that a paraconsistent logic would solve the issue of the contradictions. That is, the fact that a local change to a number, given the Peano axioms (containing the symbol for zero, $\overline{0}$, and the successor function, S), result in a number being identical to the successor of a number, and thereby violating the Peano Axioms, will not result in an explosion. The logical contradictions are indeed taken care of. However, not all of the contradictions that arise are in a sense purely logical contradictions. It remains the case that it is possible to, for example, derive that $\overline{0} = \overline{1}$.[10] This in itself could be considered to be a problem of sufficient severity, because it removes the basis of the whole system of natural numbers, i.e. the idea of succession, as discussed in the previous sections.

[10] Interestingly, in intuitionistic logic, $0 = 1$ is used as a known contradiction. Negation can, for example, be defined as $\neg A \rightarrow 0 = 1$ (Van Dalen, 2001, p. 225).

Second, with regard to a shift to a paraconsistent logic, an anti-realist of sorts might object and point out that there is an element of choice to this. Why accept the invocation of a paraconsistent logic? The need to accommodate an approach with the intention to make it so that a counterpossible is not trivially true, and to capture the explanatory role of mathematics by twiddling the mathematics itself in the antecedent, is in itself not sufficient as an argument to justify this choice. More so, even if a paraconsistent logic could constrain the contradictions, this would not be sufficient to put the concerns about the difference, or shift, in meaning aside. The point rather is whether we can speak about the same structure as the *natural* numbers after having counterfactually changed the context. Ironing out the logical contradictions appears to be only one side of the coin.

Of course, the point about resorting to a paraconsistent logic can be seen as a 'limiting case'. Jago (2014), for example, does not resort to a paraconsistent logic, and also offers an account in terms of similarity of possible and impossible worlds. A world with an alternative arithmetic in which one small deviation occurs could be said to be closer to the actual world than a world with an alternative arithmetic in which two (or more than one, in any case) deviations occur. However, such an account does indeed require the acceptance of at least a minimal overlap between antecedent and consequent (cf. the point made with regard to the first issue below). But the dispute is just as much about the acceptance of such an account. However, only if one were to accept such an account, the question of closeness becomes relevant. And in that case, it needs to be non-circularly shown that contradiction harbouring worlds can be sufficiently close to non-contradiction harbouring worlds in the relevant sense. The general point here being that this all is far from trivial.

Moving on to the first issue pertaining to the difference in meaning between the antecedent and consequent, and the need for a criterion to connect the two, another difficulty arises. Due to the change in structure of the antecedent, that is, a structure different than that of the natural numbers, the property P of the consequent, pertaining to the actual 13-year life-cycles lengths, does not find a counterpart in the antecedent. The antecedent now contains $\overline{13^*}$, whereas the consequent still contains $\overline{13}$. Given the requirement that the counterpossible should not be trivially true, there needs to be overlap between the antecedent and consequent. Therefore, the consequent should also contain $\overline{13^*}$, or at least $\overline{13^*}$ in a sense that a relevant connection exists given some property P.

But if the consequent were indeed to contain $\overline{13^*}$, then the number 13 does not occur in the explanandum anymore. However, we must not forget that the explanandum was based on the empirical data that was to be explained. And because any explanation is post hoc, the explanandum is already a given, and hence 13 has to occur in it (or perhaps a paraphrase of it in first-order logic). Consequently, to say that the physical structure must twist in line with the changes that were made to the number(s) in question (Baron et al., 2017, p. 11) raises the problem of the lost number line and the loss of the tallying procedure for the measurement of the actual life cycles. This loss was a result of the need to change the function of multiplication to multiplication*. In some sense it could be *said* that the 13 year life cycle factorizes in 6 and 2 (Baron et al., 2017, p. 11), but how is this '13' begotten? In this case, it is definitely not a tallying of the successive years, since that would already have given us the Peano numerals. This, then, is the problem this account faces. If the intention is to keep number theory the same as usual in the consequent, then, contrary to the invocation of HP in the analogy from the EIA, it is not at all easy to see what needs to be invoked in case of the counterfactual to ensure a link between the antecedent and the consequent.

8.6 Conclusion

This chapter related the consequences of a supposedly local tweak to the natural numbers in a counterpossible, as presented in Chap. 7, to relevant discussions external to the debate on counterpossibles. Section 8.1 started with a discussion on computation, where the specific place and role of symbols in computations were highlighted. These symbols, when numerals are concerned, appeared to directly specify their referent, i.e. the number they depict. This fact was picked up in Sect. 8.2. It turned out that the reason why numerals can be substituted in intensional contexts is located in the fact that they are intimately connected with de re knowledge of numbers. The suggestion was that a system of numbers is given as a whole, something which is closely connected to the idea of succession. This suggestion then functioned as a bridge to Sect. 8.3, where the canonicity of Peano numerals was discussed. The key role of the standard induction scheme, axiom (3) was mentioned. The important observation was that with it the canonicity of the Peano numerals can be deduced, and without it not. This seems to be suggestive of the fact that some form of (rudimentary) induction might already be implicit in our understanding of the natural numbers. Section 8.4 picked up on the idea of a system of numbers being given collectively, and concluded that any notational system isomorphic to the standard model of arithmetic will qualify as being about the natural numbers. The reason for this is that these notational systems all satisfy the Peano axioms.

Shifting the focus in the discussion from the antecedent of the counterfactual to the consequent, Sect. 8.5 discussed two other issues resulting from the changes made to the antecedent that have impact on the consequent. First, due to the change in structure in the antecedent, as discussed in the previous sections, the meaning of the empirically begotten consequent is now different to it. As such, there is a contextual gap between the consequent and the antecedent. It was noticed that it is not clear how to bridge this gap, or what a suggestion for it to be bridged could look like. Nonetheless, a connection is needed to avoid the counterpossible from being trivially true. In any case, since an explanation is post hoc, its content cannot be made to change following the changes of the antecedent. Second, given the contradictions that arise from the suggested local change to the number in the counterpossible, it was agreed that a shift to a paraconsistent logic would solve those. However, it was observed that not all contradictions were logical in character. Because of the fact that the successor function does not apply anymore, the equality of different numbers could still be derived. This in turn removes the basis of the whole system of (natural) numbers, which, given the discussions of the earlier sections, could be seen as a sufficiently severe problem in and of itself.

Chapter 9
Conclusion Part II

Abstract The conclusion of the second part briefly summarizes the main arguments.

The main question that was set out to be answered in Part II was what the effects of changing a mathematical antecedent of a counterfactual are. More specifically whether, given a number-theoretic counterpossible, we can be certain that we are still speaking about the same numbers in both the factual and the counterfactual case. This is important, because the statement needs to be *about* the selfsame thing. Drawing on insights external to this debate, namely about the structure of numbers, de re knowledge of numbers, and the standard interpretation of arithmetic, it appears that this is not as straightforward as it might seem. The ramifications of the intended local change to a number, such as supposing 13 to not be a prime number, result in a change of structure of the numbers.

In one case, by defining the new multiplication function 'exceptionally', this is so because the structure of the resulting theory of arithmetic is not isomorphic to the structure of the natural numbers anymore. Drawing on discussions about intended models of arithmetic, the arithmetical models about the natural numbers, it was noticed that arithmetical operations are performed through the symbols, such as Peano numerals. And intended models of arithmetic are models with recursive operations, such as addition and multiplication, satisfying the Peano axioms. Given this, it was argued that a local change to a number leads to a change in structure of the natural numbers, i.e. the new structure is not isomorphic to the standard model anymore.

In the other case, due to the special status of the standard induction scheme, Peano axiom 3, the canonicity of Peano numerals can be deduced, and without it this is not possible. This is so, because the induction axiom contains the symbol for zero, 0, and the standard successor function, S. This suggests an intimate connection between Peano numerals and de re knowledge of numbers. The counterpossible is of course concerned with numbers de re. To put it bluntly, to be a natural number is to be the member of the set with the successor function. After the local change, and the resulting change in structure, the counterfactual is not about the *natural* numbers

© The Author(s), under exclusive license to Springer Nature Switzerland AG 2025
L. A. Tump, *On Mathematical Explanations of Empirical Phenomena*,
Synthese Library 517, https://doi.org/10.1007/978-3-031-98860-8_9

anymore, but about some other numbers with a different structure. Additionally, resulting from the change in structure of the numbers mentioned in the antecedent and the numbers mentioned in the consequent of the counterpossible, it was argued that, given that the consequent now differs in structure from it, the latter does not connect to the former.

Some might argue that, in the end, none of the ramifications of the local twiddle are of real importance. For example, because the conversational context in which it is used would exclude those ramifications, in a way similar to certain irrelevant ramifications in ordinary counterfactuals are being excluded by the context. However, given the central importance of the idea of succession in our understanding of the natural numbers, i.e. in relation to the holistic aspect of systems of numbers, the analogy between the two situations cannot be taken for granted. Therefore, an account of counterfactuals with number-theoretic antecedents should take into account results both from discussions on de re knowledge of natural numbers as well as the standard interpretation of arithmetic.

These observations suggest that the way the context stops the corrosive explosion of ramifications in the case of ordinary counterfactuals, such as indeed is the case in the example of Suzy throwing the rock, might not be available in the case of number-theoretic counterfactuals, such as the suggested procedure for bringing out the explanatory contribution of mathematics in the Cicada case. To use the previously referenced metaphor of the lump in the carpet which is to be ironed out, what actually appears to happen instead of the ironing out the lump from our area of interest, is that there is no carpet left to be ironed out. Or rather, that we are left with a different carpet altogether, albeit without a lump, but a different carpet nonetheless. This is problematic for an account of explanation which is supposed to be part of an argument for realism about natural numbers.

Besides the change in the structure of the system of numbers, the emergence of a gap between the antecedent and consequent of the counterpossible was observed. If the structure of the antecedent is changed, then the consequent needs to be adjusted according to these changes. However, this is problematic, because an explanation is always post hoc. At the same time, if the intention is to bridge the gap, it is not at all clear what principle could be invoked in this case. Contrary to HP in Part I, there seems to be no clear candidate here. One suggestion for dealing with the contradictions that arise was a change in logic, i.e. some para-consistent logic. However, although a contradiction tolerant logic can certainly deal with the otherwise problematic contradictions, this alone is not enough to circumvent all problems. The reason for this is that, given the fact that the successor function does not apply anymore after making the changes to the numbers, it is still possible to derive the equality of two numbers. Being able to derive the equality of two numbers is something which already is problematic in and of itself.

Chapter 10
Epilogue

Abstract The epilogue addresses the dialectical nature of the investigations, the similarities and differences between both parts, and raises some questions related to the topics under discussion. Some initial suggestions for how to approach answers to these questions is also provided.

Having come to the end of our current investigation, some concluding remarks are in place. For instance, with regard to the dialectical nature of it, as some of you might have wondered what the employed strategy was. Generally speaking, although the investigations were for a large part logical in nature, some parts went beyond, if you like, questions of logic proper. This was perhaps most notably the case in Part II, where it was argued that solving the logical contradictions, which were introduced in the number-theoretical context of the discussed example, was not sufficient to allay all worries with regard to the counterpossible. Specifically, the resulting shift in the structure of numbers in light of the extra-mathematical explanation remained an issue. Working from the particular number-theoretic counterpossible, it transpired that such worries pertain to a wider range of examples in such contexts. In Part I there were also discussions that were not logical in nature per se, for instance when discussing the scope of an extra-mathematical explanation that reasonably could be expected given the broadly empiricist nature within which the discussion takes place. A larger scope only adds to one explanation being better than another up to a point, if begging the question is to be avoided, again, in the broadly empiricist context in which they take place. Of course, discussions around these topics, and other related topics, are far from over. However, it is quite clear that the limitations to their possible success that have come to the fore here, cannot be brushed off easily. One important reason is that the success of the strategies suggested in both Parts I and II depend on discussions outside their respective domains. And those discussions do not appear to be resolved easily, let alone soon. Additionally, there are some open question left that go beyond our current investigations, but are still related to those, and which are well suited for further research. In fact, they will probably have a direct bearing on the possible success of certain strategies.

© The Author(s), under exclusive license to Springer Nature Switzerland AG 2025

L. A. Tump, *On Mathematical Explanations of Empirical Phenomena*,
Synthese Library 517, https://doi.org/10.1007/978-3-031-98860-8_10

Before going over these open questions, let me briefly sum things up. At the very beginning it was observed that mathematics abounds in the natural sciences. Not just in physics, the science that perhaps springs to mind first, but also in others such as biology. Many theories of these sciences are formulated in so-called 'mathematicized languages'. Mathematicized languages can be thought of as languages that contain mathematical vocabulary, such as constants, functions and relation symbols. Additionally, these theories also contain mixed vocabulary in which function symbols and relation symbols are combined with some mathematical terms and some physical terms. And as we have seen, this mixed vocabulary is then used in mixed formulas whose function is to relate the mathematical terms and the physical terms. Moreover, these theories contain existentially quantified sentences of the form $\exists x \varphi$ with bound variables in φ that range over both the physical and the mathematical entities.

The scope of the existentially quantified sentences is of particular interest in the current debates, for it raises various questions as to the nature of the relation between science and mathematics, and of the ontological status of the mathematics that is used in these sentences. To put it more generally, if, as is often claimed, mathematics is indeed fundamental to the formulation of these empirical theories, indispensable even, then what does this say about the mathematics involved? A major argument in the philosophy of mathematics is the so-called 'indispensability argument'. The argument states that the theories from the natural sciences in which there is an indispensable quantification over both physical and mathematical entities are ontologically committed to the existence of those mathematical entities (and physical entities, of course). This criterion of ontological commitment was meant to show exactly, and without prejustice, what a theory says there is. However, indispensability for scientific theories is not thought to be a sufficient criterion for ontological commitment. Often, as many authors have shown, the different posits have different roles, and therefore confirmation holism seems to be too strong a requirement for determining the ontological status of some entity. Thus although it certainly seems to prevent much of the so-called 'double-talk' about what we are committed to, it does so at a price. An example of the variety of roles the different posits can have that is often discussed are idealizations, of which it is known that they are false. It seems incorrect to commit to entities of which it is known they are false, even though they are quantified over. Another example pertains to the fact that when scientific theories are tested, not all entities are considered subject of the tests. Here, some prior selection is at play, and some entities are considered to be part of the testing machinery, rather than subject of the test.

To move past the impasse, it was observed that an important role of scientific theories is to provide explanations of the natural phenomena around us. Specifically, they play a role in providing an answer to the question why something or other is the case. In some of these instances of explanations, it is argued that the explanation of the empirical phenomenon is itself mathematical. That is, the mathematics invoked is essential and integral part of the explanation of the natural phenomenon. This has led to the suggestion to restrict the set of entities under consideration for ontological commitment to only those entities that are indispensable for scientific explanations. Thus, the adjusted argument is about the role of mathematical entities in mathematical explanations of empirical phenomena, and implies that if mathematical entities play an indispensable explanatory role, then inferring their existence is justified.

In Part I of this book this adjusted, or enhanced, argument was discussed. This was done by means of an example using number theory called the 'Enhanced Indispensability Argument'. It was first argued that, perhaps somewhat surprisingly, the feasibility of the argument depends (in part) on the outcome of an another ongoing discussion on the status of a specific abstraction principle, i.e. whether Hume's Principle is an analytic or non-analytic principle. Importantly, the outcome would render the enhanced argument circular if the former were the case, that is, if the principle turns out the analytic in nature. Besides the just mentioned issue, an alternative non-mathematical explanation was presented in Part I as well. Some might argue it is somewhat limited in strength (which is of course true). However, given the fact that natural phenomena are bounded anyway, for instance as was the case with the ecological constraints in the EIA, the interpretability strategy is strong enough nonetheless to deal with all such explanations. It can do so, because it makes use of a very weak system of arithmetic, Robinson's Q, which can be interpreted in a simple theory of concatenation of non-empty strings of *a*'s and *b*'s. As such, this very elementary system should be defensible for a nominalist. Besides, it appears to be sufficiently strong as an alternative for certain applications in the natural sciences, specifically with regard to scientific explanations. That it has a more limited scope in such explanations than for instance explanations that make use of number theory, for it can only deal with numbers smaller than some number, cannot easily be used to argue against it. This is so, because invoking mathematical strength as an argument for the mathematical explanation being superior to it would result in a circular argument, because matters of scope will always be an issue with regard to explanations of natural phenomena. As it was put above, strength can only be defended up to a point.

Part II of the book focused on an argument intended to counterfactually bring out the explanatory role of mathematics with regard to the example introduced in Part I. However, given the necessary nature of mathematics, changing the properties of the mathematics in question amounts to stipulating an impossibility. Drawing on insights from discussions on the structure of numbers, de re knowledge of numbers, and the standard interpretation of arithmetic, it is argued that local changes to numbers are problematic because numbers are given collectively. Although the problem of the contradictions that appear in the counterpossible situation could be resolved by making use of some contradiction tolerant logic, for instance, the problem with allowing for local changes is actually more tenacious, because it looks like it induces a shift in context, which in turn results in a structure different from the intended one, that is, the natural numbers. For example, since the intended structure is the structure about the natural numbers, what happens after applying the suggested changes to some specific number, is that the resulting set is not closed under succession anymore. This result is quite severe, because in the example of the Periodical Cicade this means that even the simple tallying of the successive years would not be possible anymore. More so, since explanations are supposed to be post hoc, the additional question as to how the successive years can then be begotten arises. It is argued that such issues are problematic for an account of explanation which is supposed to be part of an argument for realism about the *natural numbers*.

Returning to the dialectical nature of the book, it can be said that it is, by and large, deflationary in character. It intends to show that there are problems, of logic and others, inherent to the arguments for mathematical realism in the discussed debates. Notwithstanding, the book is not only deflationary. In showing the difficulties that arise, relations to other debates thus far not considered as directly relevant to the current ones are made explicit. Most notably probably with regard to the question of the status of abstraction principles such as HP, the relative interpretability of mathematical theories, the intended models on arithmetic, or de re knowledge of numbers. Moreover, the nominalistally acceptable interpretability strategy was provided as well. This certainly could be used in other similar explanations, but that would be for further research.

Having said this, it must be stressed that, although this book takes aim at arguments for mathematical realism, it is not to be read as a defence of nominalism per se. Rather, it emphasizes the fact that nominalism is a position that cannot easily be dismissed in such discussions, particularly if these discussions take place in a broadly empiricist context.

This last mentioned remark can perhaps best be recognized in light of the marked difference between Parts I and II. In the first part induction is absent, whereas it is notably present in the second part, for instance in the discussions on de re knowledge of numbers and the intended models of arithmetic. This apparent tension between the two parts raises some interesting questions. Although, as the observant reader surely anticipates, it needs to be noted that not all arguments in Part II do require induction. Take for instance the argument in which it is shown that the stipulation that a prime number has as a factorization other than 1 and the number itself leads to a contradiction, i.e. $1 = 0$. This does not need induction indeed, as Robinson Arithmetic is sufficient for that.

However, one question is whether the arguments from de re knowledge of numbers and from the intended models of arithmetic are nominalistically acceptable despite their use of induction? An answer to this would in part depend on the specific nominalist position one holds, perhaps similar to discussions about accepting or not accepting infinitely many objects in nominalistic systems (cf. Sect. 4.2). Take for instance a formalistically inspired interpretation of numbers, where the manipulation of strings of symbols is emphasized. It must then be shown that it is possible to maintain such a position, and that induction can be formalistically acceptable, i.e. that it ultimately is about the names for the numbers.

For nominalistic positions where it is less clear whether full induction is acceptable, a question would be how much induction such a nominalist could accept if not the full induction scheme? And related to this, how much induction is needed at all in order to be able to make use of the arguments used in the context of mathematical explanations of empirical phenomena? One line of thought could be to take some form of induction in which the universal quantifier is restricted, i.e. $\forall x \leq \overline{n}$. This weaker variety should be acceptable for most nominalist positions. Additionally, it is also interesting to see what the effect of using a restricted universal quantifier is in the context of questions about the canonicity of the Peano terms. Specifically, whether it would show that Peano numerals up to n are canonical, or that the canonicity could still be shown for all Peano terms. In the latter case, the full argument used in Chap. 8 would still stand. Although a complete proof needs to be worked out for this, a first glance suggests that it is the case that in fact the canonicity of all Peano terms can be shown. The reason for this initial thought is that the standard axiom scheme gives the successor for each term that is picked out. This would be an interesting result in itself, but certainly also in the context of the topic in this book, because it would square quite well with the earlier mentioned restricted scope that is inevitable with regard to scientific explanations of natural phenomena.

Another aspect that is left for subsequent investigations regards the discussion about singling out the standard models of arithmetic. In this discussion, some of the criticism was directed at the use of Tennenbaum's theorem. This was due to its use of the notion of a recursive function in it, as it appears to already presuppose numbers. To circumvent this criticism, a move away from a theoretical notion of computability to a practical notion of computability was therefore suggested. This, however, limits the computational scope to what was called 'tractably small' numbers, and does not adjudicate over what could be computed in principle. Quite obviously, this is an entire topic in and of itself, and well beyond the scope we are currently concerned with. To mention just one interesting line of thought in relation to this problem, one could look at Robinson arithmetic, which lacks induction. By the third axiom, i.e. $\forall x((x \neq 0) \rightarrow \exists y(x = Sy))$, it gives us a sequence of elements. Note that in QT, the symbol for the successor function is defined in the additional axiom **QT6**, and can be thought of as appending an object a to an existing string x (Sect. 4.2). If Robinson Arithmetic were to be augmented by the ancestral relation, then an uninterrupted chain from any standard number that is picked out to the initial element of that chain, generally depicted by 0, is ensured. The above is the general idea underlying Peter Smith's suggestion of what he calls 'ancestral arithmetic' (Smith, 2008). Of course, Smith is concerned with Peano Arithmetic. He shows that PA^1 with the ancestral relation is conservative over PA^1. One question would then be to see whether Q with the ancestral relation is conservative over PA^1. In any case, in the context of practical computability, Robinson Arithmetic augmented with the ancestral relation omits the problems with regard to full induction. What is of importance, it seems, is that for each element that is picked out, when practically computing, there is an uninterrupted chain back to zero. The question as to how deep the notions of successor and ancestor are rooted in our understanding of numbers remains, as well as whether making changes to these relations marks a shift in the structure of the natural numbers.

L.A.T.

Bibliography

Ackerman, D. (1978). De re propositional attitudes toward integers. *Southwestern Journal of Philosophy*, *9*(2), 145–153. https://doi.org/10.5840/swjphil19789231

Azzouni, J. (2004). *Deflating existential commitment: A case for nominalism*. Oxford University Press.

Baker, A. (2003). The indispensability argument and multiple foundations for mathematics. *The Philosophical Quarterly*, *53*(210), 49–67. https://doi.org/10.1111/1467-9213.00295

Baker, A. (2005). Are there genuine mathematical explanations of physical phenomena? *Mind*, *114*(454), 223–238. https://doi.org/10.1093/mind/fzi223

Baker, A. (2009). Mathematical explanation in science. *British Journal for the Philosophy of Science*, *60*(3), 611–633. https://doi.org/10.1093/bjps/axp025

Baker, A. (2015). Mathematical explanation in biology. In *Explanation in biology, history, philosophy and theory of the life sciences* (Vol. 11). Springer. https://doi.org/10.1007/978-94-017-9822-8_10

Baker, A. (2016). Parsimony and inference to the best mathematical explanation. *Synthese*, *193*(2), 333–350. https://doi.org/10.1007/s11229-015-0723-3

Baker, A. (2017a). Mathematical spandrels. *Australasian Journal of Philosophy*, *95*(4), 779–793. https://doi.org/10.1080/00048402.2016.1262881

Baker, A. (2017b). Mathematics and explanatory generality. *Philosophia Mathematica*, *25*(2), 194–209. https://doi.org/10.1093/philmat/nkw021

Baker, A. (2021). Circularity, indispensability, and mathematical explanation in science. *Studies in History and Philosophy of Science Part A*, *88*, 156–163. https://doi.org/10.1016/j.shpsa.2021.05.017

Baker, A., & Colyvan, M. (2011). Indexing and mathematical explanation. *Philosophia Mathematica*, *19*(3), 323–334. https://doi.org/10.1093/phimat/nkr026

Balaguer, M. (1998). *Platonism and anti-platonism in mathematics*. Oxford University Press.

Balaguer, M. (2009). Fictionalism, theft, and the story of mathematics. *Philosophia Mathematica*, *17*(2), 131–162. https://doi.org/10.1093/philmat/nkn019

Bangu, S. I. (2008). Inference to the best explanation and mathematical realism. *Synthese*, *160*(1), 13–20. https://doi.org/10.1007/s11229-006-9070-8

Baron, S. (2016). The explanatory dispensability of idealizations. *Synthese*, *193*(2), 365–386. https://doi.org/10.1007/s11229-014-0517-z

Baron, S. (2020). Purely phisical explananda: Bistability in perception. In J. L. Falguera & C. Martínez-Vidal (Eds.), *Abstract objects: for and against*, Volume 422 of *Synthese Library (studies in epistemology, logic, methodology, and philosophy of science)* (pp. 17–34). Springer. https://doi.org/10.1007/978-3-030-38242-1_2

© The Author(s), under exclusive license to Springer Nature Switzerland AG 2025
L. A. Tump, *On Mathematical Explanations of Empirical Phenomena*,
Synthese Library 517, https://doi.org/10.1007/978-3-031-98860-8

Baron, S., Colyvan, M., & Ripley, D. (2017). How mathematics can make a difference. *Philosopher's Imprint*, *17*(3), 1–19.

Baron, S., Colyvan, M., & Ripley, D. (2020). A counterfactual approach to explanation in mathematics. *Philosophia Mathematica*, *28*(1), 1–34. https://doi.org/10.1093/philmat/nkz023

Benacerraf, P. (1965). What numbers could not be. *The Philosophical Review*, *74*(1), 47–73. https://doi.org/10.2307/2183530

Benacerraf, P. (1996). Recantation or any old *w*-sequence would do after all. *Philosophia Mathematica*, *4*(2), 184–189. https://doi.org/10.1093/philmat/4.2.184

Berto, F., French, R., Priest, G., & Ripley, D. (2018). Williamson on counterpossibles. *Journal of Philosophical Logic*, *47*, 693–713. https://doi.org/10.1007/s10992-017-9446-x

Berto, F., & Jago, M. (2019). *Impossible worlds*. Oxford University Press. https://doi.org/10.1093/oso/9780198812791.001.0001

Boccuni, F., & Woods, J. (2018). Structuralist neologicism. *Philosophia Mathematica*. https://doi.org/10.1093/philmat/nky017

Boolos, G. S. (1981). For every A there is a B. *Linguistic Inquiry*, *12*(3), 465–467.

Boolos, G. S. (1998). *Logic, logic, and logic*. Harvard University Press.

Boolos, G. S., Burgess, J. P., & Jeffrey, R. C. (2007). *Computability and logic* (5th ed.). Cambridge University Press.

Boyce, K. (2020). Mathematical application and the no confirmation thesis. *Analysis*, *1*, 11–20. https://doi.org/10.1093/analys/anz021

Brogaard, B., & Salerno, J. (2013). Remarks on counterpossibles. *Synthese*, *190*(4), 639–660. https://doi.org/10.1007/s11229-012-0196-6

Bueno, O. (2001). Logicism revisited. *Principia: An International Journal of Epistemology*, *5*(1–2), 99–124.

Bueno, O. (2012). An easy road to nominalism. *Mind*, *121*(484), 967–982. https://doi.org/10.1093/mind/fzs114

Bueno, O. (2016). An anti-realist account of the application of mathematics. *Analysis*, *173*, 2591–2604. https://doi.org/10.1007/s11098-016-0670-y

Bueno, O. (2023). Dispensing with facts, substances, and structures. *American Philosophical Quarterly*, *60*(1), 49–61. https://doi.org/10.5406/21521123.60.1.05

Bueno, O., & Colyvan, M. (2011). An inferential conception of the application of mathematics. *Noûs*, *45*(2), 345–374. https://doi.org/10.1111/j.1468-0068.2010.00772.x

Bueno, O., & French, S. (2012). Can mathematics explain physical phenomena? *British Journal for the Philosophy of Science*, *63*(1), 85–113. https://doi.org/10.1093/bjps/axr017

Burgess, J. P. (2005). *Fixing Frege*. Princeton University Press.

Burgess, J. P., & Rosen, G. (1997). *A Subject with no object: Strategies for nominalistic interpretation of mathematics*. Oxford University Press.

Busch, J., & Morrison, J. (2016). Should scientific realists be platonists? *Synthese*, *193*(2), 435–449. https://doi.org/10.1007/s11229-015-0676-6

Button, T., & Smith, P. (2012). The philosophical significance of Tennenbaum's theorem. *Philosophia Mathematica*, *20*(1), 114–121. https://doi.org/10.1093/philmat/nkr031

Carnap, R. (1998/1966). The value of laws: Explanation and prediction. In M. Curd & J. A. Cover (Eds.), *Philosophy of science, the central issues* (pp. 678–684). W.W. Norton & Company, Inc.

Cartwright, N. (1980). Do the laws of physics state the facts? *Pacific Philosophical Quarterly*, *61*(1–2), 75–84. https://doi.org/10.1111/j.1468-0114.1980.tb00005.x

Cartwright, N. (1983). *How the laws of physics lie*. Oxford University Press.

Colyvan, M. (2001). *The indispensability of mathematics*. Oxford University Press.

Colyvan, M. (2002). Mathematics and aesthetic considerations in science. *Mind*, *111*(441), 69–74.

Colyvan, M. (2010). There is no easy road to nominalism. *Mind*, *119*(474), 285–306. https://doi.org/10.1093/mind/fzq014

Colyvan, M. (2012). *An introduction to the philosophy of mathematics*. Cambridge University Press.

Colyvan, M. (2013). Road work ahead: Heavy machinery on the easy road. *Mind*, *121*(484), 1031–1046. https://doi.org/10.1093/mind/fzt014

Curd, M., & Cover, J. A. (1998). *Philosophy of science, the central issues*. W.W. Norton & Company, Inc.

Daly, C., & Langford, S. (2009). Mathematical explanation and indispensability arguments. *The Philosophical Quarterly*, *59*(237), 641–658. https://doi.org/10.1111/j.1467-9213.2008.601.x

Damnjanovic, Z. (2017). Mutual interpretability of Robinson Arithmetic and Adjunctive Set Theory with Extensionality. *Bulletin of Symbolic Logic*, *23*(4), 381–404. https://doi.org/10.1017/bsl.2017.30

Darnell, E., & Thomas Bolduc, A. (2018). Is Hume's principle analytic? *Synthese*. https://doi.org/10.1007/s11229-018-01988-8

Dijksterhuis, E. J. (1961/1950). *The mechanization of the world picture*. Oxford Clarendon Press.

Douven, I. (2017). Abduction. In E. N. Zalta (Ed.), *The Stanford encyclopedia of philosophy* (Summer 2017 ed.). Metaphysics Research Lab, Stanford University.

Duhem, P. (1954/1906a). *The aim and structure of physical theory*. Princeton University Press.

Duhem, P. (1998/1906b). Physical theory and experiment. In M. Curd & J. A. Cover (Eds.), *Philosophy of science, the central issues* (pp. 257–279). W.W. Norton & Company, Inc.

Einstein, A. (1921). Geometry and experience. In *CPAE* (English translation) (Vol. 7, pp. 208–222). Princeton University Press.

Enderton, H. B. (2001). *A mathematical introduction to logic*. Academic Press.

Ferreira, F., & Ferreira, G. (2013). Interpretability in Robinson's Q. *Bulletin of Symbolic Logic*, *19*(3), 289–317. https://doi.org/10.2178/bsl.1903010

Field, H. (1980). *Science without numbers*. Princeton University Press.

Field, H. (1984). Frege's conception of numbers as objects by Crispin Wright. *Canadian Journal of Philosophy*, *14*, 637–662.

Field, H. (1989). *Realism, mathematics and modality*. Basil Blackwell Ltd.

Field, H. (1992). A nominalistic proof of the conservativeness of set theory. *Journal of Philosophical Logic*, *21*(2), 111–123.

Frege, G. (1893). *Grundgesetze der arithmetik, begriffsschriftlich abgeleitet*. Verlag Hermann Pohle.

Gijsbers, V. (2017). A quasi-interventionist theory of mathematical explanation. *Logique et Analyse*, *60*(237), 47–66. https://doi.org/10.2143/LEA.237.0.3197358

Goodman, N., & Quine, W. V. O. (1947). Steps toward a constructive nominalism. *Journal of Symbolic Logic*, *12*(4), 105–122. https://doi.org/10.2307/2266485

Grattan-Guinness, I. (2008). Solving Wigner's mystery: the reasonable (though perhaps limited) effectiveness of mathematics in the natural sciences. *The Mathematical Intelligencer*, *30*, 7–17. https://doi.org/10.1007/BF02985373

Hafner, J., & Mancosu, P. (2005). The varieties of mathematical explanation. In P. Mancosu, K. F. Jørgensen, & Pedersen, S. A. (Eds.), *Visualization, explanation and reasoning styles in mathematics*, Volume 327 of *Synthese Library (studies in epistemology, logic, methodology, and philosophy of science)* (pp. 215–250). Springer. https://doi.org/10.1007/1-4020-3335-4_9

Halbach, V., & Horsten, L. (2005). Computational structuralism. *Philosophia Mathematica*, *13*(2), 174–186. https://doi.org/10.1093/philmat/nki021

Halvorson, H. (2019). *The logic in philosophy of science*. Cambridge University Press.

Hempel, C. G. (1962). Explanation in science and in history. In R. G. Colodney (Ed.), *Frontiers of science and philosophy* (pp. 9–33). University of Pittsburgh Press.

Hempel, C. G. (1965). *Aspects of scientific explanation and other essays in the philosophy of science*. The Free Press.

Hempel, C. G., & Oppenheim, P. (1948). Studies in the logic of explanation. *Philosophy of Science*, *15*(2), 135–175. https://doi.org/10.1086/286983

Heylen, J. (2021/2014). The epistemic significance of numerals. *Synthese*, *198*, 1019–1045. https://doi.org/10.1007/s11229-014-0542-y

Hintikka, J., & Halonen, I. (1999). Semantics and pragmatics for why-questions. In *Inquiry as inquiry: A logic of scientific discovery. Jaakko Hintikka Selected Papers* (Vol. 5, pp. 183–204). Springer. https://doi.org/10.1007/978-94-015-9313-7_9

Horsten, L. (2012). Vom Zählen zu den Zahlen: on the relation between computation and arithmetical structuralism. *Pilosophia Mathematica*, *20*(3), 275–288. https://doi.org/10.1093/philmat/nkr039

Hume, D. (2007/1748). *An enquiry concerning human understanding*. Oxford University Press.

Hunt, J. (2016). Indispensability and the problem of compatible explanations: A reply to 'should scientific realists be platonists?'. *Synthese*, *193*(2), 451–467. https://doi.org/10.1007/s11229-015-0667-7

Hylton, P. (2007). *Quine*. Routledge.

Ito, H., Kakishima, S., Uehara, T., Morita, S., Koyama, T., Sota, T., Cooley, J. R., & Yoshimura, J. (2015). Evolution of periodicity in periodical cicadas. *Scientific Reports*, *5*(14094). https://doi.org/10.1038/srep14094

Jago, M. (2014). *The impossible: An essay on hyperintensionality*. Oxford University Press.

Jenny, M. (2018). Counterpossibles in science: The case of relative computability. *Noûs*, *52*(3), 530–560. https://doi.org/10.1111/nous.12177

Kaplan, D. (1968). Quantifying in. *Synthese*, *19*(1/2), 178–214. https://doi.org/10.1007/BF00568057

Ketland, J. (1998). *The mathematicization of nature*. Ph. D. thesis, London School of Economics.

Kitcher, P. (1989). Explanatory unification and the causal structure of the world. In P. Kitcher & W. C. Salmon (Eds.), *Scientific explanation* (pp. 410–505). University of Minnesota Press.

Kline, M. (1972). *Mathematical thought from ancient to modern times*. Oxford University Press.

Knowles, R., & Saatsi, J. (2019). Mathematics and explanatory generality: Nothing but cognitive salience. *Erkenntnis*, 1119–1137. https://doi.org/10.1007/s10670-019-00146-x

Kripke, S. A. (1980). *Naming and necessity*. Harvard University Press.

Kripke, S. A. (2011). *Philosophical troubles: Collected papers, volume I*. Oxford University Press.

Ladyman, J., Douven, I., Horsten, L., & Van Fraassen, B. (1997). A defence of Van Fraassen's critique of abductive inference: Reply to Psillos. *The Philosophical Quarterly*, *47*(188), 305–321. https://doi.org/10.1111/1467-9213.00061

Landau, E. (1958). *Elementary number theory*. Chelsea Publishing Company.

Lange, M. (2012). What makes a scientific explanation distinctively mathematical. *British Journal for the Philosophy of Science*, *64*(3), 485–511. https://doi.org/10.1093/bjps/axs012

Lange, M. (2017). *Because without cause: Non-causal explanations in science and mathematics*. Oxford University Press.

Lawrence, R. (2021). Giving the value of a variable. *Kriterion - Journal of Philosophy*. https://doi.org/10.1515/krt-2021-0007

Lehman-Ziebarth, N., Heideman, P. P., Shapiro, R. A., Stoddart, S. L., Hsiao, C. C. L., Stephenson, G. R., Milewski, P. A., & Irvine, A. R. (2005). Evolution of periodicity in periodical cicadas. *Ecology*, *86*(12), 3200–3211. https://doi.org/10.1890/04-1615

Leng, M. (2005). Mathematical explanation. In C. Cellucci & D. Gillies (Eds.), *Mathematical reasoning, heuristics and the development of mathematics* (pp. 167–189). King's College Publications.

Leng, M. (2010). *Mathematics and reality*. Oxford University Press.

Lewis, D. (1973a). Causation. *The Journal of Philosophy*, *70*(17), 556–567. https://doi.org/10.2307/2025310

Lewis, D. (1973b). *Counterfactuals*. Blackwell Publishers.

Lewis, D. (1979). Counterfactual dependence and time's arrow. *Noûs*, *13*(4), 455–476. https://doi.org/10.2307/2215339

Lewis, D. (1983). New work for a theory of universals. *Australasian Journal of Philosophy*, *61*, 343–377. https://doi.org/10.1080/00048408312341131

Lewis, D. (1986). Causal explanation. In *Philosophical papers* (Vol. 2, pp. 214–240). Oxford University Press.

Liggins, D. (2008). Quine, Putnam, and the 'Quine-Putnam' indispensability argument. *Erkenntnis*, *68*(1), 113–127. https://doi.org/10.1007/s10670-007-9081-y

Lipton, P. (2004). *Inference to the best explanation* (2nd ed.). International library of philosophy. Routledge.

Lyon, A., & Colyvan, M. (2008). The explanatory power of phase spaces. *Philosophia Mathematica*, *16*(2), 227–243. https://doi.org/10.1093/philmat/nkm025

Maddy, P. (1992). Indispensability and practice. *Journal of Philosophy*, *89*(6), 275–289. https://doi.org/10.2307/2026712

Maddy, P. (1997). *Naturalism in mathematics*. Clarendon Press.

Maddy, P. (2008). How applied mathematics became pure. *The Review of Symbolic Logic*, *1*(1), 16–41. https://doi.org/10.1017/S1755020308080027

Marcus, R. (2014a). The holistic presumptions of the indispensability argument. *Synthese*, *191*, 3575–3594. https://doi.org/10.1007/s11229-014-0481-7

Marcus, R. (2014b). How not to enhance the indispensability argument. *Philosophia Mathematica*, *22*(3), 345–360. https://doi.org/10.1093/philmat/nku004

Melia, J. (2000). Weaseling away the indispensability argument. *Mind*, *109*(435), 455–480. https://doi.org/10.1093/mind/109.435.455

Melia, J. (2002). Response to Colyvan. *Mind*, *111*, 75–79. https://doi.org/10.1093/mind/111.441.75

Molinini, D., Pataut, F., & Sereni, A. (2016). Indispensability and explanation: An overview and introduction. *Synthese*, *193*, 317–332. https://doi.org/10.1007/s11229-015-0998-4

Morrison, M. (2000). *Unifying scientific theories: Physical concepts and mathematical structures*. Cambridge University Press.

Niebergall, K.-G. (2011). Mereology. In R. Pettigrew & L. Horsten (Eds.), *The continuum companion to philosophical logic* (pp. 271–298). Bloomsbury.

Nolan, D. (1997). Impossible worlds: a modest approach. *Notre Dame Journal of Formal Logic*, *38*(4). https://doi.org/10.1305/ndjfl/1039540769

Okasha, S. (2000). Van Fraassen's critique of inference to the best explanation. *Studies in History and Philosophy of Science*, *31*(4), 691–710. https://doi.org/10.1016/S0039-3681(00)00016-9

Panza, M., & Sereni, A. (2016). The varieties of indispensability arguments. *Synthese*. https://doi.org/10.1007/s11229-015-0977-9

Pincock, C. (2012). *Mathematics and scientific representation*. Oxford studies in philosophy of science. Oxford University Press.

Pincock, C. (2015). Abstract explanations in science. *British Journal for the Philosophy of Science*, *66*(4), 857–882. https://doi.org/10.1093/bjps/axu016

Price, D. M. (2017). *Application and ontology in mathematics: A defence of fictionalism*. Ph. D. thesis, University of York.

Priest, G. (2016). Thinking the impossible. *Philosophical Studies*, *173*(10), 2649–2662. https://doi.org/10.1007/s11098-016-0668-5

Psillos, S. (1996). On Van Fraassen's critique of abductive reasoning. *The Philosophical Quarterly*, *46*(182), 31–47. https://doi.org/10.2307/2956303

Putnam, H. (1971). *Philosophy of logic*. Allen & Unwin.

Putnam, H. (1975). What is mathematical truth? In *Mathematics, matter and method* (Vol. 1, pp. 60–78). Cambridge University Press.

Putnam, H. (2012/2006). Indispensability arguments in the philosophy of mathematics. In M. D. Caro & D. Macarthur (Eds.), *Philosophy in an age of science: Physics, mathematics, and skepticism* (Chap. 9, pp. 181–201). Harvard University Press.

Quine, W. V. O. (1948). On what there is. *Review of Metaphysics*, *2*(5), 21–38.

Quine, W. V. O. (1951). Two dogmas of empiricism. *The Philosophical Review*, *60*, 20–43. https://doi.org/10.2307/2266637

Quine, W. V. O. (1956). Quantifiers and propositional attitudes. *The Journal of Philosophy*, *53*(5), 177–187. https://doi.org/10.2307/2022451

Quine, W. V. O. (1960a). Carnap and logical truth. *Synthese*, *12*(4), 350–374. https://doi.org/10.1007/BF00485423

Quine, W. V. O. (1960b). *Word and object*. The MIT Press.

Quine, W. V. O. (1961). *From a logical point of view* (2nd revised (1961) ed.). Harvard University Press.

Quine, W. V. O. (1961/1953). On what there is. In *From a logical point of view* (2nd ed., pp. 1–19). Harvard University Press.

Quine, W. V. O. (1981a). Success and limits of mathematization. In *Theories and things* (pp. 148–155). Harvard University Press.

Quine, W. V. O. (1981b). Things and their place in theories. In *Theories and things* (pp. 1–23). Harvard University Press.

Quine, W. V. O. (1986). *Philosophy of logic* (2nd ed.). Harvard University Press.

Railton, P. (1978). A deductive-nomological model of probabilistic explanation. *Philosophy of Science*, *45*, 206–226. https://doi.org/10.1086/288797

Rescorla, M. (2007). Church's thesis and the conceptual analysis of computability. *Notre Dame Journal of Formal Logic*, *48*(2), 253–280. https://doi.org/10.1305/ndjfl/1179323267

Rescorla, M. (2015). The representational foundations of computation. *Philosophia Mathematica*, *23*(3), 338–366. https://doi.org/10.1093/philmat/nkv009

Resnik, M. D. (1997). *Mathematics as a science of patterns*. Clarendon Press. https://doi.org/10.1111/1467-9213.00267

Resnik, M. D., & Kushner, D. (1987). Explanation, independence and realism in mathematics. *The British Journal for the Philosophy of Science*, *38*(2), 141–158. https://doi.org/10.1093/bjps/38.2.141

Reutlinger, A. (2016). Is there a monist theory of causal and noncausal explanations? The counterfactual theory of scientific explanation. *Philosophy of Science*, *83*, 733–745. https://doi.org/10.1086/687859

Reutlinger, A. (2017). Explanation beyond causation? New directions in the philosophy of scientific explanation. *Philosophy Compass*, *12*(2). https://doi.org/10.1111/phc3.12395

Russell, B. (1905). On denoting. *Mind*, *14*(56), 479–493. https://doi.org/10.1093/mind/XIV.4.479

Saatsi, J. (2011). The enhanced indispensability argument: representational versus explanatory role of mathematics in science. *The British Journal for the Philosophy of Science*, *62*(1), 143–154. https://doi.org/10.1093/bjps/axq029

Saatsi, J. (2016). On the 'indispensable explanatory role' of mathematics. *Mind*, *125*(500), 1045–1070. https://doi.org/10.1093/mind/fzv175

Saatsi, J., & Pexton, M. (2013). Reassessing Woodward's account of explanation: regularities, counterfactuals, and noncausal explanations. *Philosophy of Science*, *80*(5), 613–624. https://doi.org/10.1086/673899

Salmon, W. C. (1971/1970). Statistical explanation. In W. C. Salmon (Ed.), *Statistical explanation & statistical relevance* (pp. 29–87). University of Pittsburgh Press.

Salmon, W. C. (1984). *Scientific explanation and the causal structure of the world*. Princeton University Press.

Sandborg, D. (1998). Mathematical explanation and the theory of why-questions. *The British Journal for the Philosophy of Science*, *49*(4), 603–624. https://doi.org/10.1093/bjps/49.4.603

Sellars, W. (1963). *Science, perception and reality*. Ridgeview Publishing Company.

Shapiro, S. (1982). Acceptable notation. *Notre Dame Journal of Formal Logic*, *23*(1), 14–20. https://doi.org/10.1305/ndjfl/1093883561

Shapiro, S. (1983). Mathematics and reality. *Philosophy of Science*, *50*(4), 523–548. https://doi.org/10.1086/289138

Shapiro, S. (1997). *Philosophy of mathematics: Structure and ontology*. Oxford University Press.

Shapiro, S. (2000). *Thinking about mathematics: The philosophy of mathematics*. Oxford University Press.

Shapiro, S. (2017). Computing with numbers and other non-syntactic things: *de re* knowledge of abstract objects. *Philosophia Mathematica*, *25*(2), 268–281. https://doi.org/10.1093/philmat/nkx009

Sklar, L. (1974). *Space, time, and spacetime*. University of California Press.

Smid, J. (2020). The logic behind Quine's criterion of ontological commitment. *European Journal of Philosophy*. https://doi.org/10.1111/ejop.12534

Smith, P. (2008). Ancestral arithmetic and Isaacson's Thesis. *Analysis*, *68*(1), 1–10.

Stalnaker, R. C. (1968). A theory of conditionals. In N. Rescher (Ed.), *Studies in logical theory*, American philosophical quarterly monographs 2 (pp. 98–112). Basil Blackwell Publisher.
Steiner, M. (1978a). Mathematical explanation. *Philosophical Studies*, *34*(2), 135–151. https://doi.org/10.1007/BF00354494
Steiner, M. (1978b). Mathematics, explanation, and scientific knowledge. *Noûs*, *12*(1), 17–28. https://doi.org/10.2307/2214652
Steiner, M. (2011). Kripke on logicism, Wittgenstein, and *De Re* beliefs about numbers. In A. Berger (Ed.), *Saul Kripke*. Cambridge University Press.
Strevens, M. (2006). Scientific explanation. In D. Borchert (Ed.), *The encyclopedia of philosophy* (2nd ed.). Macmillen.
Strevens, M. (2008). *Depth: An account of scientific explanation*. Harvard University Press.
Tallant, J. (2013). Optimus Prime: paraphrasing prime number talk. *Synthese*, *190*(12), 2065–2083. https://doi.org/10.1007/s11229-011-9959-8
Tarski, A. (1936). The concept of truth in formalized languages. In A. Tarski (Ed.), *Logic, semantics, metamathematics* (pp. 152–278). Oxford University Press.
Tarski, A., Mostowski, A., & Robinson, R. M. (1953). *Undecidable theories*. North-Holland Publishing Company.
Taylor, J. R. (2005). *Classical mechanics*. University Science Books.
Van Dalen, D. (2001). Intuitionistic logic. In L. Goble (Ed.), *The Blackwell guide to philosophical logic* (pp. 224–257). Blackwell Publishers.
Van Dalen, D. (2004). *Logic and structure* (4th ed.). Springer-Verlag.
Van Fraassen, B. C. (1980). *The scientific image*. Clarendon Press.
Van Fraassen, B. C. (1989). *Laws and symmetry*. Clarendon Press.
Walters, L. (2016). Possible world semantics and true-true counterfactuals. *Pacific Philosophical Quarterly*, *97*(3), 322–346. https://doi.org/10.1111/papq.12067
Weber, E., & Verhoeven, L. (2002). Explanatory proofs in mathematics. *Logique et Analyse*, *45*(197–180), 299–307.
Wigner, E. P. (1960). The unreasonable effectiveness of mathematics in the natural sciences. Richard Courant lecture in mathematical sciences delivered at New York University, May 11, 1959. *Communications on Pure and Applied Mathematics*, *13*, 1–14. https://doi.org/10.1002/cpa.3160130102
Williamson, T. (2007). *The philosophy of philosophy*. Blackwell Publishing Ltd.
Williamson, T. (2018). Counterpossibles. *Topoi*, *37*(3), 257–368. https://doi.org/10.1007/s11245-016-9424-x
Wittgenstein, L. (1967/1956). *Remarks on the foundations of mathematics*. The MIT Press.
Woodward, J. (2003). *Making things happen: A theory of causal explanation*. Oxford University Press.
Wright, C. (1999). Is Hume's principle analytic? *Notre Dame Journal of Formal Logic*, *40*(1), 6–30. https://doi.org/10.1305/ndjfl/1039096303
Ylikoski, P., & Kuorikoski, J. (2010). Dissecting explanatory power. *Philosophical Studies*, *148*, 201–219. https://doi.org/10.1007/s11098-008-9324-z

The manufacturer's authorised representative in the EU is Springer Nature Customer Service Centre GmbH, Europaplatz 3, 69115 Heidelberg, Germany. If you have any concerns regarding our products, please contact ProductSafety@springernature.com

Printed and bound by CPI Group (UK) Ltd, Croydon, CR0 4YY
30/09/2026
02225513-0006